Bunch and Oil Analysis of Oil Palm

A Manual

Techniques in Plantation Science Series

Series editors:

Brian P. Forster, Lead Scientist, Verdant Bioscience, Indonesia
Peter D.S. Caligari, Science Strategy Executive Director, Verdant Bioscience, Indonesia

About the series:

A series of manuals covering techniques in plantation science that form the essential underlying needs to carry out plantation science.

The series reflects the expertise in Verdant Bioscience that underlies the plantation science activities carried out at the Verdant Plantation Science Centre at Timbang Deli, Deli Serdang, North Sumatra, Indonesia.

Titles available:

1. Crossing in Oil Palm: A Manual – Umi Setiawati, Baihaqi Sitepu, Fazrin Nur, Brian P. Forster and Sylvester Dery
2. Seed Production in Oil Palm: A Manual – Eddy S. Kelanaputra, Stephen P.C. Nelson, Umi Setiawati, Baihaqi Sitepu, Fazrin Nur, Brian P. Forster and Abdul R. Purba
3. Nursery Screening for *Ganoderma* Response in Oil Palm Seedlings: A Manual – Miranti Rahmaningsih, Ike Virdiana, Syamsul Bahri, Yassier Anwar, Brian P. Forster and Frédéric Breton
4. Mutation Breeding in Oil Palm: A Manual – Fazrin Nur, Brian P. Forster, Samuel A. Osei, Samuel Amiteye, Jennifer Ciomas, Soeranto Hoeman and Ljupcho Jankuloski
5. Bunch and Oil Analysis of Oil Palm: A Manual – Pujo Widodo, Fazrin Nur, Evi Nafisah, Brian P. Forster and Hasrul Abdi Hasibuan

Bunch and Oil Analysis of Oil Palm
A Manual

Pujo Widodo
Verdant Bioscience, Indonesia

Fazrin Nur
Verdant Bioscience, Indonesia

Evi Nafisah
Verdant Bioscience, Indonesia

Brian P. Forster
Verdant Bioscience, Indonesia

Hasrul Abdi Hasibuan
Indonesian Oil Palm Research Institute, Indonesia

CABI is a trading name of CAB International

CABI	CABI
Nosworthy Way	745 Atlantic Avenue
Wallingford	8th Floor
Oxfordshire OX10 8DE	Boston, MA 02111
UK	USA
Tel: +44 (0)1491 832111	Tel: +1 (617)682-9015
Fax: +44 (0)1491 833508	E-mail: cabi-nao@cabi.org
E-mail: info@cabi.org	
Website: www.cabi.org	

© Pujo Widodo, Fazrin Nur, Evi Nafisah, Brian P. Forster and Hasrul Abdi Hasibuan 2019. All rights reserved. No part of this publication may be reproduced in any form or by any means, electronically, mechanically, by photocopying, recording or otherwise, without the prior permission of the copyright owners.

A catalogue record for this book is available from the British Library, London, UK.

Library of Congress Cataloging-in-Publication Data

Names: Widodo, Pujo, author. | Nur, Fazrin, author.
Title: Bunch and oil analysis of oil palm : a manual / Pujo Widodo, Fazrin Nur, Evi Navisah, Brian P. Forster, Hasrul Abdi Hasibuan.
Description: Oxfordshire, UK ; Boston, MA : CABI, [2019] | Series: Techniques in plantation science | Includes bibliographical references and index.
Identifiers: LCCN 2018059732 (print) | LCCN 2019010341 (ebook) | ISBN 9781789241372 (ePDF) | ISBN 9781789241389 (ePub) | ISBN 9781789241365 (paperback : alk. paper)
Subjects: LCSH: Oil palm--Analysis.
Classification: LCC SB299.P3 (ebook) | LCC SB299.P3 W53 2019 (print) | DDC 633.8/51--dc23
LC record available at https://lccn.loc.gov/2018059732

ISBN-13: 978 1 78924 136 5 (paperback)
        9781789241372 (ePDF)
        9781789241389 (ePub)

Commissioning editor: Rebecca Stubbs
Editorial assistant: Emma McCann
Production editor: James Bishop

Typeset by SPi, Pondicherry, India
Printed and bound in the UK by Severn, Gloucester

Series Foreword – Techniques in Plantation Science

Verdant Bioscience, Singapore (VBS) is a company established in October 2013 with a vision to develop high-yielding, high-quality planting material in oil palm and rubber through the application of sound practices based on scientific innovation in plant breeding. The approach is to fuse traditional breeding strategies with the latest methods in biotechnology. These techniques are integrated with expertise and the application of sustainable aspects of agronomy and crop protection, alongside information and imaging technology which not only find relevance in direct aspects of plantation practice but also in selection within the breeding programme. When high-yielding planting material is allied with efficient plantation practices, it leads to what may be termed 'intensive sustainable' production. At the same time, the quality of new products is refined to give more specialized uses alongside more commodity-based oil production, thus meeting the market demands of the modern world community, but with a minimal harmful footprint. An essential ingredient in all this is having sound and practical protocols and techniques to allow the realization of the strategies that are envisaged.

To achieve its aims, VBS acquired an Indonesian company called PT Timbang Deli Indonesia, with an estate of over 970 hectares of land at Timbang Deli, Deli Serdang, North Sumatra, Indonesia, and the group works under the name of 'Verdant'. A central part of this estate, which will be used for important plant nurseries and field trials, is the development of the Verdant Plantation Science Centre (VPSC), to which the operational staff moved in October 2016. A seed production and marketing facility is now established at VPSC for commercial seed sales and the processing of seed from breeding programmes. The centre comprises specialized laboratories in cell biology, genomics, tissue culture, pollen, soil DNA, plant and soil nutrition, bunch and oil, agronomy and crop protection. Field facilities include extensive nurseries, seed gardens and trials (trial sites are also located at various locations across Indonesia). It is the aim of the company to use its existing and rapidly developing intellectual property (IP) to develop superior cultivars that not only have outstanding yield but are also resistant to both biotic and abiotic stresses, while at the same time meeting new market demands. Verdant not only develops and supplies superior planting materials but also supports its customers and growers with a package of services and advice in fertilizer recommendations and crop protection. This is all part of a central mission to promote green, eco-friendly agriculture.

Brian P. Forster and Peter D.S. Caligari
Lead Scientist and Science Strategy Executive Director
Verdant Bioscience

Contents

Acknowledgements	ix
Preface	xi

1	**Introduction**	3
	1.1 Oil Palm Production Facts and Figures	3
	1.2 Palm Oil Fruit Morphology and Oil Composition	4
	1.3 Palm Oil Production	5
	1.4 Methods in Bunch and Oil Yield Analyses	7
	1.5 Laboratory Set-up and Work Flow	11
	References	13
2	**Health and Safety Considerations**	17
	2.1 Health and Safety in the Field	17
	2.2 Health and Safety in the Laboratory	18
	References	19
3	**Bunch Sampling**	21
	3.1 Steps in Bunch Sampling	21
	3.2 Tools and Equipment	25
4	**Bunch Physical Analysis**	29
	4.1 Steps in Bunch Physical Analysis	29
	4.2 Tools and Equipment	39
	4.3 Materials	41
	Reference	43
5	**Fruit Sampling**	45
	5.1 Steps in Fruit Sampling	45
	5.2 Tools and Equipment	48
6	**Nut Analysis**	53
	6.1 Steps in Nut Analysis	53
	6.2 Tools and Equipment	56
	Reference	58

7	**Oil Analysis**	**61**
	7.1 Steps in Oil Analysis	61
	7.2 Tools and Equipment	68
	7.3 Materials	70
	Reference	73
8	**Recording, Calculations and Data Checks**	**75**
	8.1 Recording	75
	8.2 Calculations	76
	8.3 Tools and Equipment	79
	8.4 Data Checks and Quality Control	79
	References	83
Index		**85**

Acknowledgements

The authors are grateful to all the breeding and biotechnology teams of Verdant for sharing their knowledge and providing helpful advice in preparing this manual.

Brian P. Forster and Peter D.S. Caligari
Lead Scientist and Science Strategy Executive Director
Verdant Bioscience

Preface

As noted in the foreword to this series, a central objective in Verdant's mission is to develop better, more productive and more sustainable cultivars of oil palm, rubber and other plantation crops, through plant breeding. The higher yielding the planting material, the less land that is needed to achieve a specific level of production in terms of oil per hectare. Essential to this objective is the ability to measure yield in breeding lines and to compare this with standard yields of current varieties. Commercial estates are also keenly interested to know how productive their plantations are, and how efficient their oil mills are in extracting oil from oil palm bunches. Thus, physical bunch and oil analysis is of interest and increasingly bunch and oil analysis laboratories are being established at oil palm mills to measure yield and particularly oil extraction rates. Physical bunch and oil analyses are central to Verdant's breeding programmes in selecting and developing high yielding cultivars. Protocols developed for physical bunch and oil analyses form the basis for this manual. Our target audiences are growers, millers, students and researchers in agriculture, plant breeders and end users interested in the practicalities of measuring yield in oil palm for commercial production and breeding.

Brian P. Forster and Peter D.S. Caligari
Lead Scientist and Science Strategy Executive Director
Verdant Bioscience

Introduction

Abstract

A brief history of bunch analysis of the oil palm crop is provided. This includes bunch components (stalk, spikelets, fruits, mesocarp, nuts, shell and kernels) and oil yield analysis. These parameters are used by oil palm breeders to analyse the oil yield potentials (e.g. quantity and quality comparison of bunches, between bunches within an oil palm, between oil palms or between different progenies). Obtaining high fresh fruit bunch (FFB) production in oil palm plantations per hectare area per year, and high oil extraction rates from the bunch, are the main goals of growers to produce maximal oil yield. Oil extraction rates determined in the laboratory are useful in this respect, in monitoring and verifying extraction rates in oil mills.

1.1 Oil Palm Production Facts and Figures

Oil palm has the Latin name *Elaeis guineensis* Jacq; the genus name is derived from the Greek *elaion*, meaning 'oil', and the species name indicates its origin in West Africa. The crop was discovered by travellers to Africa in the 15th century, but the first plantings in Indonesia, which led to its rise as the world's pre-eminent oil crop, did not occur until the late 19th century. Large scale plantations were established in the early 20th century in both Africa and South East Asia as interest in the crop developed. The first plantations were composed of Dura palms which have thick-shelled fruits (Fig. 1.1). In the 1920s the first crosses were made in deliberate attempts to improve the crop through plant breeding, and in the 1950s–60s the more productive Tenera types (a result of crossing Dura with Pisifera) took over as the favoured commercial material both in Africa and South East Asia. Tenera genotypes are thin-shelled and have thick oil-bearing fruit flesh (mesocarp) with a 30% greater oil yield compared to Duras. Oil is also obtained from the kernel.

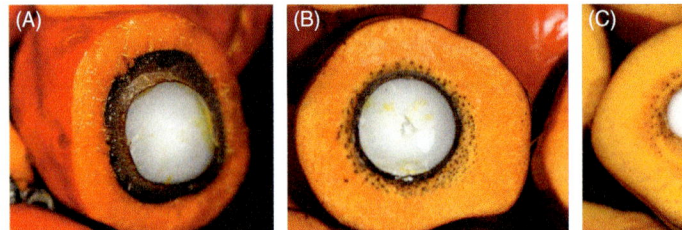

Fig. 1.1. Fruit types of oil palm: A) Dura, thick-shelled; B) Tenera, thin-shelled with a fibre ring around the shell and a thick mesocarp (fruit flesh); C) Pisifera, no shell but with traces of a fibre ring around the kernel. In addition to the orange fruit flesh, the kernel (white) is a valuable source of high quality oil.

A comprehensive review of the oil palm crop is given by Corley and Tinker (2015). Oil palm is grown in the humid tropics between 20° latitudes north and south of the equator and covers over 8.5 million hectares worldwide. The crop is highly profitable and grown both on large scale plantations and by smallholders (Sayer *et al.*, 2012). It should be noted that of all the oil producing crops exploited at present, oil palm is by far the most efficient on an oil volume per hectare basis.

1.2 Palm Oil Fruit Morphology and Oil Composition

The oil palm fruit consists of an outer pericarp, an outer skin (exocarp), an oily flesh (mesocarp) and a shell (endocarp). The pericarp organs are maternal and derived from the carpel of the mother palm. The kernel is composed mainly of endosperm that supports the growth of the enclosed embryo. The endosperm and embryo are products of fertilization. Crude palm oil (CPO) is extracted from the outer mesocarp and crude palm kernel oil (CPKO) is extracted from the endosperm.

The mesocarp of oil palm fruits is made up of the three major components: oil, water and fibre. Oil palm fruits provide both CPO and palm kernel oil (PKO), extracted from the fruit flesh (mesocarp) and kernel (endosperm), respectively (Fig 1.2). Oil palm is by far the highest oil yielding crop, averaging 3–4 tonnes of mesocarp oil per hectare per year in the major palm oil producing countries (Wahid *et al.*, 2005). Fatty acids in CPO consist of palmitic acid (44%), oleic acid (39%), linoleic acid (11%), stearic acid (5%) and other fatty acids (Siew, 2002), and CPO is a major source of pro-vitamin A and vitamin E (Barcelos *et al.*, 2015). Vitamin E content in CPO ranges between 600 ppm and 1,000 ppm (Choo *et al.*, 1995). Carotene and sterols are minor components of CPO (Tenera) and the average carotene content in Deli Dura is about 500 ppm (Corley and Tinker, 2015). The Malaysian standard for carotene content in CPO is 474–689 ppm (Malaysian Standard MS 814, 2007). PKO is a high quality oil containing lauric acid (up to 50%),

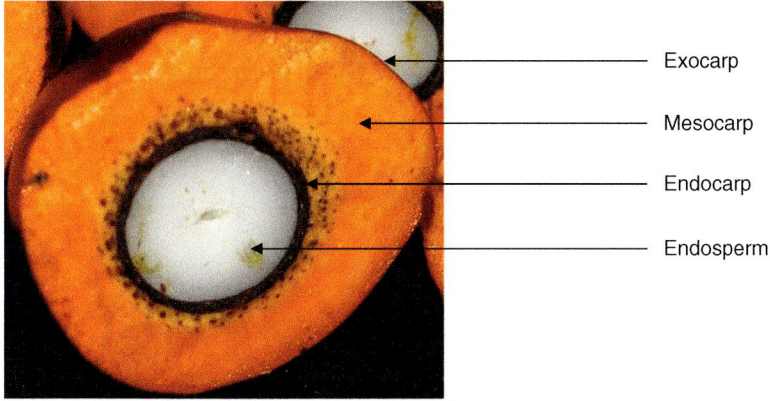

Fig. 1.2. Palm oil fruit morphology (Tenera, thin-shelled).

myristic acid (17%), palmitic acid (8%), oleic acid (13%) and other essential fatty acids (Siew, 2001).

The shell thickness characteristic (see Fig. 1.1) of Dura is thick-shelled: the shell is 2–8 mm thick, occasionally less, 35–65% mesocarp/fruit, no fibre ring surrounding the nut in cross-section; Tenera fruits are thin-shelled: shell 0.5–4 mm, 55–96% mesocarp/fruit and a fibre ring is present; Pisifera are shell-less and usually female sterile (Corley and Tinker, 2015).

Recent comparative studies of the characteristics of FFB yield of palm oil and palm kernel oil of Dura, Pisifera and Tenera genotypes in North Sumatra, Indonesia (Basyuni *et al.*, 2017) show:

- Oil percentages of mesocarp: 56% in Tenera, 49% in Dura and 61% in Pisifera fruits.
- CPO percentages: 24% in Tenera, 23% in Dura and 26% in Pisifera fruits.

1.3 Palm Oil Production

Harvesting should allow the collection of all ripe bunches in a plantation without damaging the fruit (or the palm) to obtain optimal oil content and oil quality. Methods of harvesting are usually written down in standard operating procedures (SOPs). The SOPs of harvesting should cover the specific interval time between rounds of harvesting a block of palm trees in a plantation as well as observation notes. The maximum time interval is about 10 days, which should prevent any bunch becoming over-ripe. A study comparing harvesting intervals from 10 to 15 days was carried out by Mohanaraj and Donough (2016) and showed 10 days to be the best harvest interval. Harvesting observations should be made on developing fruit bunches – e.g. fruit colour changes (black to red for nigrescens genotypes; green to orange for virescens genotypes), number of loose fruits (detached fruits on the ground) before and after harvesting, fruit ripeness, etc.

Fruit colour helps the harvester to identify the maturity stage of fruit bunches (either nigrescens or virescens type – see Fig. 1.3). Over-ripe fruit will have bright orange and yellow colours in nigrescens and virescens types respectively. Ripe fruits will have a yellow/orange or red/orange colour; under-ripe fruits will have a reddish black colour or green colour; and unripe fruits will have a mostly black or green colour in nigrescens and virescens fruit respectively. The number of loose fruits (fruits found on the

Fig. 1.3. Unripe and ripe fruits A) Unripe nigrescens; B) Ripe nigrescens; C) Unripe virescens; D) Ripe virescens.

ground) before harvesting is normally 1–5 or 5–10 fruits, depending on the estate SOPs (this corresponds to the ripeness and quality of the fruit bunch). Under-ripe and unripe bunches will diminish the oil quality due to higher free fatty acids and lower oil yield. Also, over-ripe fruit bunches will lose oil during field handling and transportation to the mill. In some cases, the harvesting SOPs also count the number of loose fruits after harvesting (less than five loose fruit/kg bunch weight), minimum weight of bunch (minimum weight is normally 3 kg) and stalk length (less than 2 cm). Ripe fruits have optimum oil extraction rates and good quality. Woittiez *et al.* (2016) list the following characteristics for ripe fruits:

- Yellow/orange/red colour for nigrescens types.
- Yellow/orange coloured fruits for virescens types.
- Good bunch appearance (no pests or diseases).
- High fertility (bunch is full of fruits).
- Fruits are soft and oil drips out when fruit is cut.

Ripe fruit bunches are harvested continually and sent to local mills for oil extraction and high scale production of CPO and CPKO (Fig. 1.4). The main CPO producing countries are Indonesia and Malaysia, which supply 55% and 30% of the total global production, respectively. These two countries produce 63.82 million tonnes (MnT) per year; Indonesia provides 54 and Malaysia 36% of total CPO exports worldwide, equivalent to 46.74 MnT. The largest importers are India (20%), followed by Europe (16%) and China (11%), with a total import equal to 46.71 MnT (Oil World, 2017).

Most palm oil is used in the food industry and is a major source of lipids (Wahid *et al.*, 2005). Palm oil is used mainly for edible purposes, whereas palm kernel oil has wide applications in the oleochemical industry (Sambanthamurthi *et al.*, 2000). The downstream oleochemical industry produces a wide range of food and care products, and uses palm oil in the production of frying oils, margarines, shortenings, filling fats, detergents, soaps, surfactants, emulsifiers, biodiesels, pharmaceutical products, etc.

1.4 Methods in Bunch and Oil Yield Analyses

Bunch and yield analyses are used to determine the physical properties and yield of oil in oil palm fruit bunches after harvesting. Yield components assessed in these analyses are: bunch sample weight; fruit type (Dura, Tenera and Pisifera, see Fig. 1.1); stalk weight; total number of spikelets in the bunch; percentage of fertile fruits in the bunch (F/B); number of parthenocarpic fruits in the bunch; number of undeveloped fruits in the bunch; fruit weight (FWT); percentage of mesocarp in the fruits (M/F); mean nut weight (MNW); percentage of kernel in the fruit (K/F); percentage of shell in the fruit (S/F); percentage of kernel in the bunch (K/B); moisture content (%) of

Fig. 1.4. Harvesting: A) Cutting ripe bunch from the palm; B) Bunch collection; C) Bunch transport; D) Bunch arrival at mill (with kind permission from the Bukit Maradja Estate and Bukit Maradja mill of PT. Tolan Tiga Indonesia).

the mesocarp (fruit flesh); percentage of oil in wet mesocarp samples (O/WM); percentage of oil in dried mesocarp samples (O/DM); and percentage of mesocarp oil in the bunch (O/B).

The fresh fruit bunch should be transported to the bunch analysis laboratory as soon as possible after harvesting to prevent any changes in its physical components and quality, i.e. moisture loss and altered free fatty acid content. To minimize these changes, it is recommended that the freshly harvested fruit bunch is placed into a gunny sack or other bag that will help reduce moisture losses and fruit damage during transport to the bunch analysis laboratory.

Once the harvested bunches are received in the laboratory, they are fermented overnight using ethepon solution (an ethylene producer that promotes fruit ripening). Ethepon treatment can be done either by injection into the stalk, or by spraying the solution all over the bunch surface. After overnight incubation at room temperature (25–30°C) in the dark, bunches are weighed and then chopped in order to separate the spikelets from the stalk. The stalk is weighed to calculate the percentage of stalk to the bunch. A fruit bunch typically contains about 100 spikelets, and the spikelets should be counted to obtain the spikelet number per bunch. Then a specific amount of spikelets are taken as a sample (details are given in Chapter 4).

There are two methods of spikelet sampling. The first is to take a random sample of a certain weight or number; the second takes into account uneven fruit set in different parts of a bunch and the variation in bunch components in different regions of the bunch (Rao *et al.*, 1983). In the first method, all spikelets are separated and thoroughly mixed in a mixing box, which is then tipped over into two empty boxes standing side by side in such a way that each box contains a similar sample size of at least 5 kg of spikelets (see Chapter 3). Both spikelet box samples are weighed. One (approximately 5 kg) is retained for analysis and the other sample is discarded. (Blaak *et al.*, 1963). This method has been modified by Rao *et al.* (1983) as follows:

> "The spikelets are thoroughly mixed on the table with a shovel. A sample of spikelets is then taken by pulling the moveable surround forward so that the spikelets fall off the front of the table into a collecting box. The collecting box is designed in such a way that a strip of spikelets from front to back along one side of the table is collected, the rest are discarded".

In a second method (Ooi and Tam, 1975), three regions are distinguished for each bunch: apical, middle and basal. Each of these regions is subdivided into the part which is adpressed to the trunk and the part that is adpressed to the frond, thus giving six specific sample regions per bunch.

Spikelet samples are weighed before proceeding to the peeling step. The fruits are separated into fertile fruits, parthenocarpic (sterile) fruits and under-developed fruits after the peeling step (removal of the fruits from the spikelets).

The fertile fruits are weighed to calculate the percentage of fertile fruits in the bunch (F/B), while the parthenocarpic fruits and under-developed fruits are counted. Bruised and chopped fruits are removed from the fertile fruit samples and then taken to the random mixing box. Blaak *et al.* (1963) recommended a sample size of 500 g, but other laboratories commonly use 250 g, which is represented by 25–30 fruits. In this manual we use a 30-fruit sample, which is weighed to calculate the mean fruit weight.

The individual fruits of the sample are scraped with a knife to separate the fresh mesocarp flesh from the nut. The fresh mesocarp is weighed to calculate the percentage of mesocarp in the fruits (M/F), while the nuts are weighed and counted to calculate the mean nut weight (MNW). The percentage of kernel in the fruit (K/F), percentage of shell in the fruit (S/F) and percentage of kernel in the bunch (K/B) are calculated after the nuts are oven dried (105°C for at least seven hours). After heat treatment the nuts are cracked and the nut shell and kernel are weighed.

The percentage moisture content of the mesocarp is determined by drying the fresh mesocarp in an oven (at 105°C for 24 hours) followed by weighing the dry mesocarp. For oil extraction purposes, the dried mesocarp should be ground and sieved (2 mm mesh size). The mesocarp is dried again for 30 minutes at 105°C and about 8 g is placed in envelopes made from Whatman filter paper.

The envelopes may be stored in a desiccator or used directly for oil extraction. Envelopes are stacked inside a soxhlet extractor chamber and the oil extracted using n-hexane as the solvent in a reflux system for 24 hours, or until the solvent becomes clear. The percentage of oil in the dry mesocarp sample (O/DM), the percentage of oil in the wet (fresh) mesocarp (O/WM) and the percentage of mesocarp oil in the bunch (O/B) are calculated from the weights of the mesocarp before and after extraction.

Computation formulae for bunch components are after Junaidah *et al.* (2011).

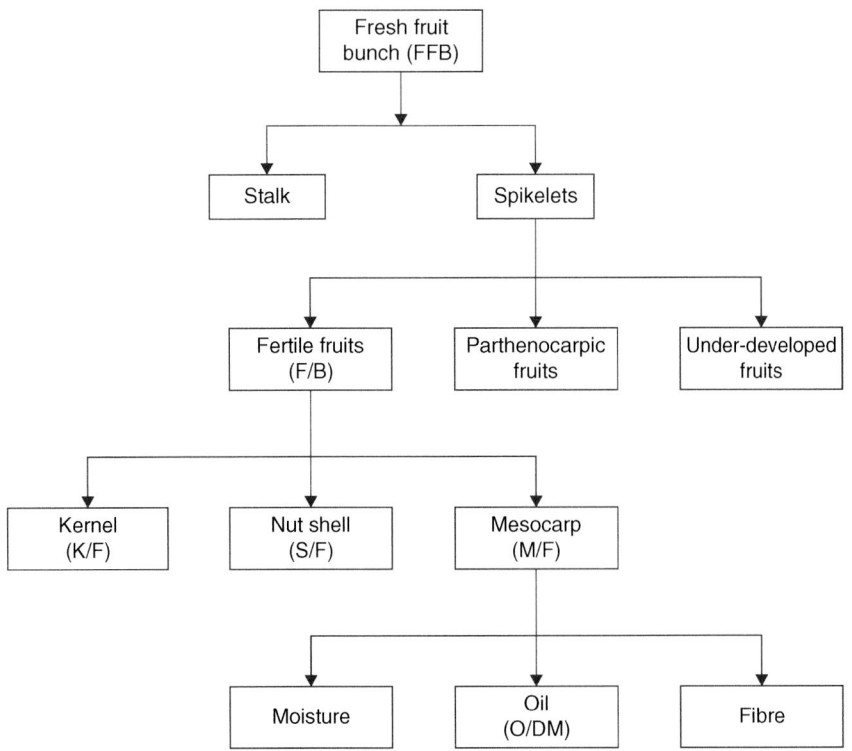

Fig. 1.5. Steps in oil yield analysis starting from a fresh fruit bunch.

1.5 Laboratory Set-up and Work Flow

Bunch and oil activities are conducted in the laboratory and consist of two main parts: bunch laboratory to analyse basic bunch characteristics (bunch, stalk, spikelets, fruits, nuts and kernel); and oil laboratory to determine oil content in the dry mesocarp sample. A basic design of a bunch and oil laboratory is given below, showing the flow of samples (Fig. 1.6) and waste (Fig. 1.7) to and from the various work stations.

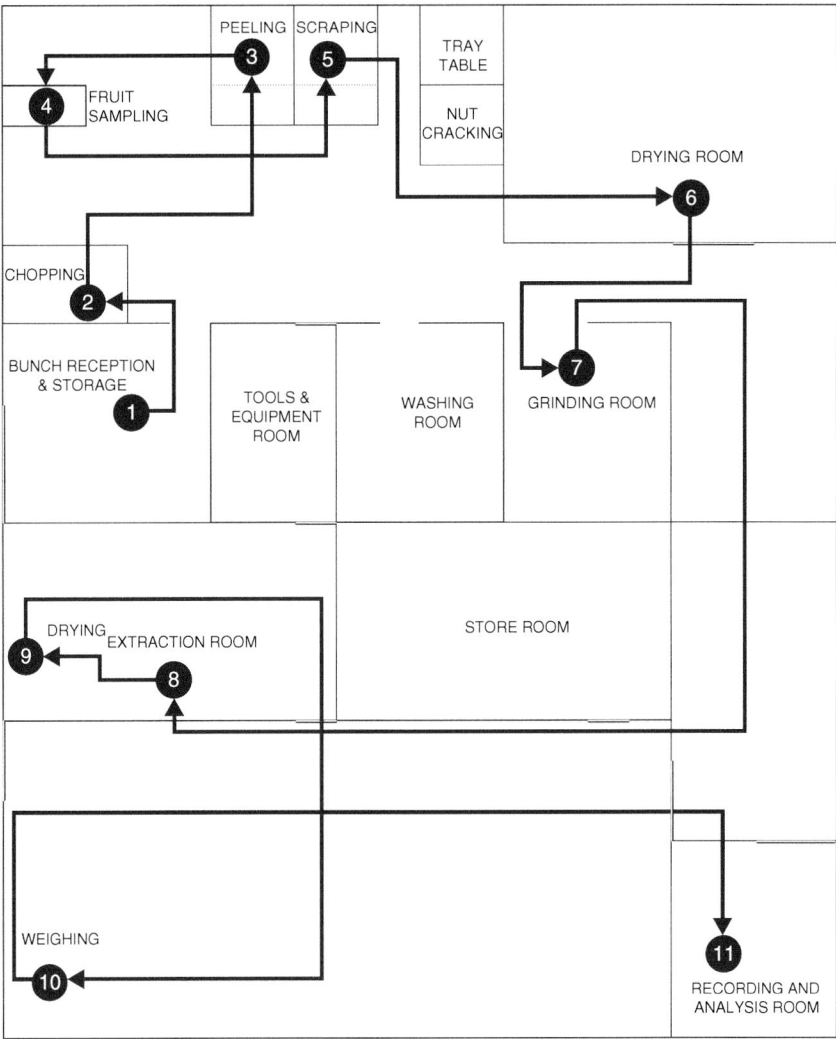

Fig. 1.6. Flow of materials and samples through the laboratory.

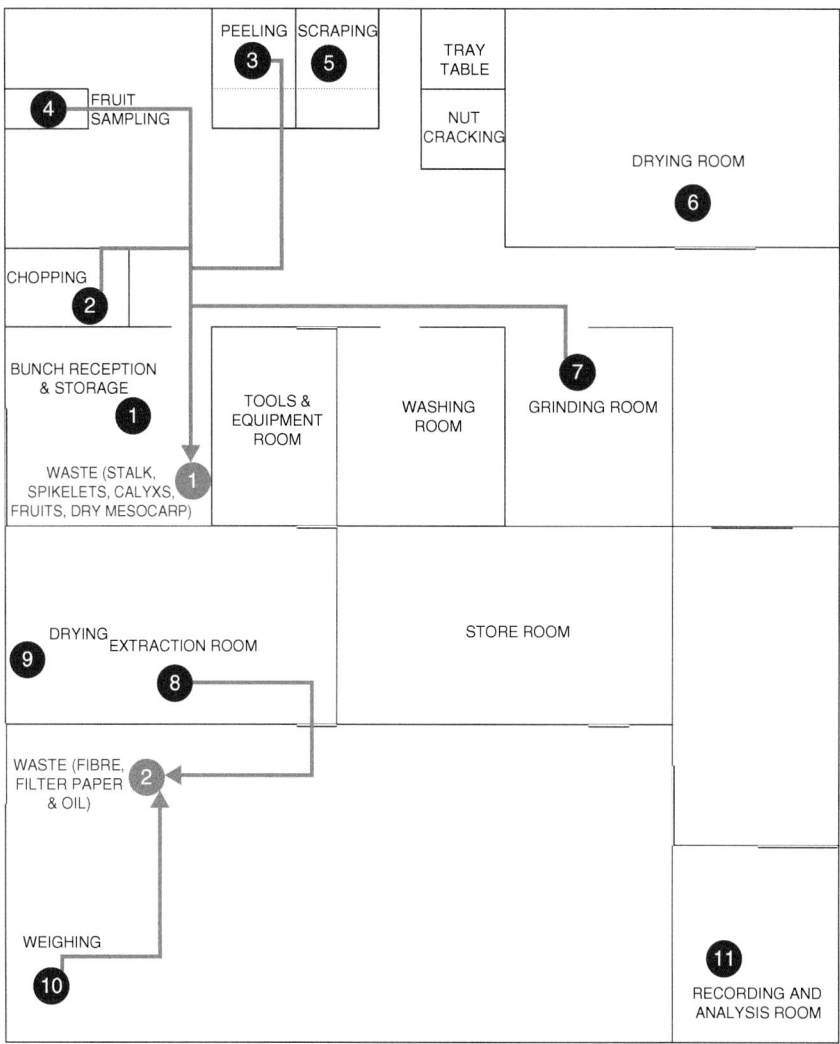

Fig. 1.7. Flow of waste through the laboratory.

Harvested bunches from the field are delivered to the bunch reception section, followed by the bunch characterization section: bunch weighing, bunch chopping (to separate the spikelets from the stalk), stalk weighing, spikelet sampling, spikelet weighing, fruit peeling and fruit weighing; fruit analysis section: fruit sampling, fruit scraping, fresh mesocarp weighing, mesocarp drying and dry mesocarp weighing; nut analysis section: nut counting, nut drying, nut cracking, kernel weighing and kernel counting; oil analysis section: mesocarp grinding, ground mesocarp sieving, drying, ground mesocarp weighing, oil extraction and fibre weighing; and recording section.

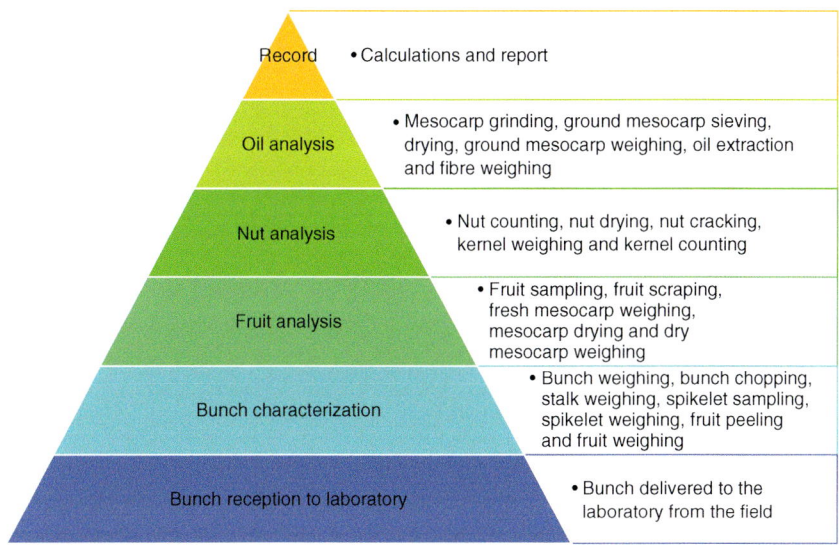

Fig. 1.8. Bunch and oil laboratory procedures.

Blaak *et al.* (1963) divided the analysis laboratory into six sections: a base section for the reception of bunches; a section for fruit form determination, weighing the bunches, separation of spikelets from the stalk and spikelet sampling of the bunch; a section for sampling fruit from the spikelets, weighing empty spikelets and weighing the fruit; a section for sampling fruit, scraping fruit, weighing, drying and cracking nuts, and weighing kernels; oil analysis section; and a record section (as summarized in Fig. 1.8).

References

Barcelos, E., Rios, S.D.A., Cunha, R.N.V., Lopes, R., Motoike, S.Y. *et al.* (2015) Oil palm natural diversity and the potential for yield improvement. *Frontiers in Plant Science*, Volume 6(190): pp. 1–16.

Basyuni, M., Amri, N., Putri, L.A.P., Syahputra, I. and Arifiyanto, D. (2017) Characteristics of fresh fruit bunch yield and the physicochemical qualities of palm oil during storage in North Sumatra, Indonesia. *Indonesian Journal of Chemistry*, Volume 17(2): Indonesia, pp. 182–190.

Blaak, G., Sparnaaij, L.D. and Menendez, T. (1963) Breeding and inheritance in oil palm: part II. Methods of bunch quality analysis. *Journal of West African Institute for Oil Palm Research*, Volume 4(14): Benin, pp. 145–155.

Choo, Y.M., Ma, A.N. and Yap, S.C. (1995) Carotenes, vitamin E and sterols in oils from E. *guineensis*, E. *oleifera* and their hybrids. *Palm Oil Developments*, Volume 27: PORIM, Malaysia, pp. 1–8.

Corley, R.H.V. and Tinker, P.B. (2015) *The Oil Palm*, fifth edition. Wiley Blackwell, UK, p. 28.

Junaidah, J., Kushairi, A., Jones, B., Kho, L.E., Isa, Z.A. and Rusmin, J. (2011) Innovation for oil extraction method using NMR in bunch analysis. *International Seminar on Breeding for Sustainability in Oil Palm*, 18 November, ISOPB & MPOB, Kuala Lumpur, Malaysia, pp. 1–18.

Malaysian Standard MS 814 (second revision) (2007) Specification for crude palm oil, Department of Standards Malaysia, Kuala Lumpur, Malaysia, pp. 1–6.

Mohanaraj, S.N. and Donough, C.R. (2016) Harvesting practices for maximum yield in oil palm: result from a re-assessment at IJM plantations, Sabah. *Oil Palm Bulletin*, Volume 72 (May): MPOB, Malaysia, pp. 32–37.

Oil World (2017) *Oil World Monthly Report No. 2*, Volume 60: 13 January, ISTA Mielke GmbH, Hamburg, pp. 1–18.

Ooi, S.C. and Tam, T.K. (1975) The determination of the within bunch components of oil yield in the oil palm (*Elaeis guineensis* Jacq.) I. Pattern of variation. *MARDI Res. Bulletin*, Volume 3: pp. 44–47.

Rao, V., Soh, A.C., Corley, R.H.V., Lee, C.H., Rajanaidu, N. *et al.* (1983) A critical re-examination of the method of bunch quality analysis in oil palm breeding. PORIM *Occasional Paper* no. 9, MPOB, Malaysia, 7–8.

Sambanthamurthi, R., Kalyana, S. and Tan, Y. (2000) Chemistry and biochemistry of palm oil. *Progress in Lipid Research*, Volume 39: Malaysia, pp. 507–558.

Sayer, J., Ghazoul, J., Nelson, P. and Boedhihartono, A.K. (2012) Oil palm expansion transforms tropical landscapes and livelihoods. *Global Food Security*, Volume 1(2): pp. 114–119.

Siew, W.L. (2001) Crystallisation and melting behaviour of palm kernel oil and related products by differential scanning calorimetry. *European Journal of Lipid Science Technology*, Volume 103: pp. 729–734.

Siew, W.L. (2002) Palm oil. In: Gunstone, F.D. (ed.) *Vegetable Oils in Food Technology: Composition, Properties and Uses*. Wiley-Blackwell Publishing, New Jersey, pp. 58–97.

Wahid, M.B., Abdullah, S.N.A. and Henson, I.E. (2005) Oil palm – achievements and potential. *Plant Production Science Journal*, Volume 8(3): pp. 289–297.

Woittiez, L.S., Haryono, S., Turhina, S., Dani, H., Dukan, T.P. and Smit, H. (2016) *Smallholder oil palm handbook module 2: Harvesting, grading, transport, second edition*. Wageningen University, Wageningen, and SNV International Development Organisation, The Hague, pp. 8–10.

Health and Safety Considerations 2

Abstract

Bunch and oil analysis involves activities in the field and the laboratory. All field and laboratory operations should adhere to standard health and safety protocols. These may vary according to local requirements and standards. Some equipment will also come with instructions on proper use, which may involve training, including health and safety issues. Failure to abide by these can result in accidents and personal injury (serious and minor); neglect of health and safety issues may also incur penalties such as fines or cessation of field and laboratory activities. Guidelines in health and safety issues relating to bunch sampling in oil palm are given below.

2.1 Health and Safety in the Field

Bunch sampling is carried out in the field and a major health and safety issue is the height of palm trees being sampled. The clothing and equipment needed for harvesting the oil palm bunch includes:

- Boots to protect from rough terrain, spines and ground animals.
- Sensible clothing, including hats to protect from sunlight.
- Cloth gloves and helmet.
- Sharp chisel or sickle.

Other general considerations are:

1. Training in the use of equipment and tools: workers should be fully trained in the use of a chisel and/or sickle used during harvesting bunches in the field.
2. Standard operating procedures, e.g. harvesting SOPs.
3. Awareness of sharp spines of oil palms.

4. Awareness of dangerous/harmful insects and animals.
5. Dangers of working alone.
6. Emergency procedures, first-aid box.

2.2 Health and Safety in the Laboratory

General guidelines for working at the bench are given by Barker (2005). Yield analysis of oil palm bunches requires various laboratory procedures, especially in the chopping of bunch stalks with an axe, scraping the mesocarp from the nut and oil extraction from dry mesocarp samples using a hazardous solvent (hexane). Basic laboratory safety applies.

1. Put on a laboratory coat before entering the laboratory and remove it when leaving the laboratory. This provides protection for yourself and the samples you are working with. It also protects people outside the laboratory from contamination by laboratory materials.
2. Use specific personal protective equipment (PPE) during the handling of chemicals in the laboratory, e.g. mask, eye protection and gloves.
3. Be aware of emergency procedures: fire-fighting, emergency signals, emergency exits, emergency phone numbers, location of fire extinguishers, emergency meeting points, first-aid box/first-aiders and local medical facilities.
4. Be aware of hazards relating to the chemicals used in the laboratory and their safety data sheet (SDS): chemical identification; hazard(s) identification; composition/information on ingredients; first-aid measures; fire-fighting measures; accidental release measures; handling and storage; exposure controls/personal protection; physical and chemical properties; stability and reactivity; toxicological information; ecological information; disposal considerations; transport information; regulatory information; and other relevant information.
5. Workers should be fully trained in the use of sharp tools, e.g. use of electrical drill to drill the bunches; use of an axe to chop the bunches to separate the stalks and spikelets; and use of a knife to scrape the mesocarp from the nut.
6. Be careful when working with glass, especially when handling the glassware, e.g. glass beaker, boiling flask and soxhlet, etc.
7. Take care when injecting ethephon into bunch stalk. Refer to the SDS during the handling of this chemical (Bayer CropScience, 2009).
8. Be aware of the solvent air exposure during solvent extraction. Solvent extraction should be carried out in a specialized room with ventilation and an air extraction system above the soxhlets. Be careful when using flammable solvent in case of fire. Do not expose the solvent in the air near a heat source or electrical plug.
9. Be especially careful when handling n-hexane during oil extraction. Refer to SDS. Hexane is a dangerous chemical (Merck Millipore, 2016).

Hazard statements of n-hexane:

- H225 Highly flammable liquid and vapour.
- H304 May be fatal if swallowed and/or enters airways.
- H315 Causes skin irritation.
- H336 May cause drowsiness or dizziness.
- H361 Suspected of damaging fertility. Suspected of damaging the unborn child.
- H373 May cause damage to organs (nervous system) through prolonged or repeated exposure if inhaled.
- H411 Toxic to aquatic life with long-lasting effects.

Precautionary statements of n-hexane:

- Prevention
  P210 Keep away from heat, hot surfaces, sparks, open flames and other ignition sources. No smoking.
  P240 Ground/bond container and receiving equipment.
  P273 Avoid release into the environment.
- Response
  P301 + P330 + P331 IF SWALLOWED: rinse mouth. Do NOT induce vomiting.
  P302 + P352 IF ON SKIN: wash with plenty of soap and water.
  P314 Get medical advice/attention if you feel unwell.
- Storage
  P403 + P233 Store in a well-ventilated place. Keep container tightly closed.

10. Be aware of SOPs that have been developed for your laboratory, or which should be developed, e.g. solvent/waste storage and disposal which include the storage condition, storage container, storage room or stored separately, maximum storage volume, etc.

References

Barker, K. (2005) *At the Bench: A Laboratory Navigator*. Cold Spring Harbor Press, New York, NY.

Bayer CropScience (2009) Ethrel® Liquid Plant Growth Regulator Safety Data Sheet (SDS). Version 2.0 CDN 102000004255 Revision 01/12/2017, pp. 1–11.

Merck Millipore (2016) n-Hexane Safety Data Sheet (SDS). 104374_SDS_Hexane EN. Revision 31.10.2016, Version 4.4, pp. 1–18.

Bunch Sampling

3

Abstract

Bunch sampling is the harvesting of fresh fruit bunches (FFBs) from oil palm trees. This is done according to standard operating procedures (SOPs), with specific timed intervals between each round of harvesting. During bunch sampling, observations are conducted on the status of the bunches, e.g. fruit colour, number of loose fruits before and after harvesting on the ground, etc. These will correspond to the recoverable oil yield and quality – e.g. free fatty acid content. Good practices are essential to maximize and standardize data capture for oil yield and quality.

3.1 Steps in Bunch Sampling

Step 1

Identify a ripe bunch from a selected individual palm based on leaf sampling unit (LSU) – a standard sampling area often used by an agronomist or a trial description. The bunch is ripe when the fruit colour is reddish (in nigrescens fruit type) or orange (in virescens type) (Fig. 3.1).

Fig. 3.1. Step 1: Find ripe fruit bunches of: A) Nigrescens type (red); B) Virescens type (orange).

Step 2

Count the loose fruit on the ground (shed from the ripe bunch) and record the number of loose fruits before harvesting the ripe bunch. There should be at least one loose fruit, but no more than five, on the ground (though this can vary according to estate practices). Record the number of loose fruits (Fig. 3.2).

Fig. 3.2. Step 2: Loose fruit (one) on the ground shed from the ripe bunch. Count the loose fruit.

Step 3

Cut the bunch from the palm using a chisel. This is a specialized skill for which training is required. Cut the frond underlying the bunch as close as possible to the frond base, then cut the bunch stalk to allow the bunch to fall to the ground (Fig. 3.3).

Fig. 3.3. Step 3: Cutting the bunch from the palm using a chisel.

Step 4

Count the number of loose fruits (on the ground) after harvesting. Estimate the bunch weight. The number of loose fruits after harvesting should be less than five loose fruit/kg bunch weight. If above this, the bunch sample should be discarded and another bunch sampled. Record the number of loose fruits (Fig. 3.4).

Fig. 3.4. Step 4: Counting the number of loose fruits after harvesting.

Step 5

Prepare the label which will show the trial number, LSU number, field number and row/palm number. This identifies the sample (Fig. 3.5).

Estate   :
Field Number   :
Trial Number   :
LSU Number   :
Row/Palm Number:

Fig. 3.5. Step 5: Label to identify sample.

Step 6

Place the harvested bunch along with its loose fruits into a gunny sack, along with the label inside the gunny sack, and tie with plastic rope. Collect all the bunch samples and transport to the bunch laboratory (Fig. 3.6). A three-wheeled motorcycle or estate dump truck are commonly used. Make sure that no loose fruits are lost during transportation.

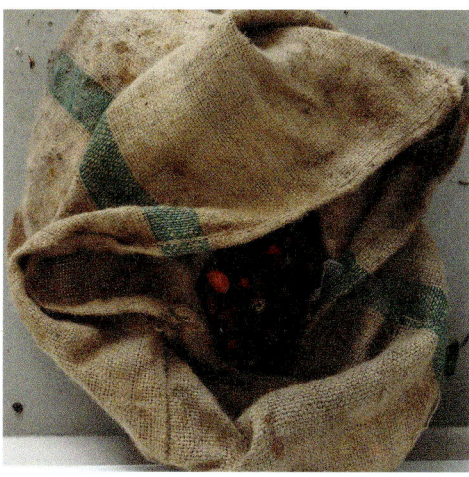

Fig. 3.6. Step 6: Place the bunch and loose fruits into a gunny sack and transport to the laboratory.

3.2 Tools and Equipment

- Boots – used to protect feet from sharp spines.
- Chisel – used to harvest the bunch from oil palm tree.
- Gunny sack – used to collect the harvested bunch and fruits.
- Plastic rope – used to tie the gunny sack with the bunch sample inside.
- Label – used to label the bunch sample.
- Ballpoint pen – used to write label information for the bunch sample.
- Three-wheeled motorcycle/estate truck – used for transporting the bunch samples from the field to the laboratory.
- Recording form – used for recording the number of loose fruits before harvesting and number of loose fruits after harvesting.

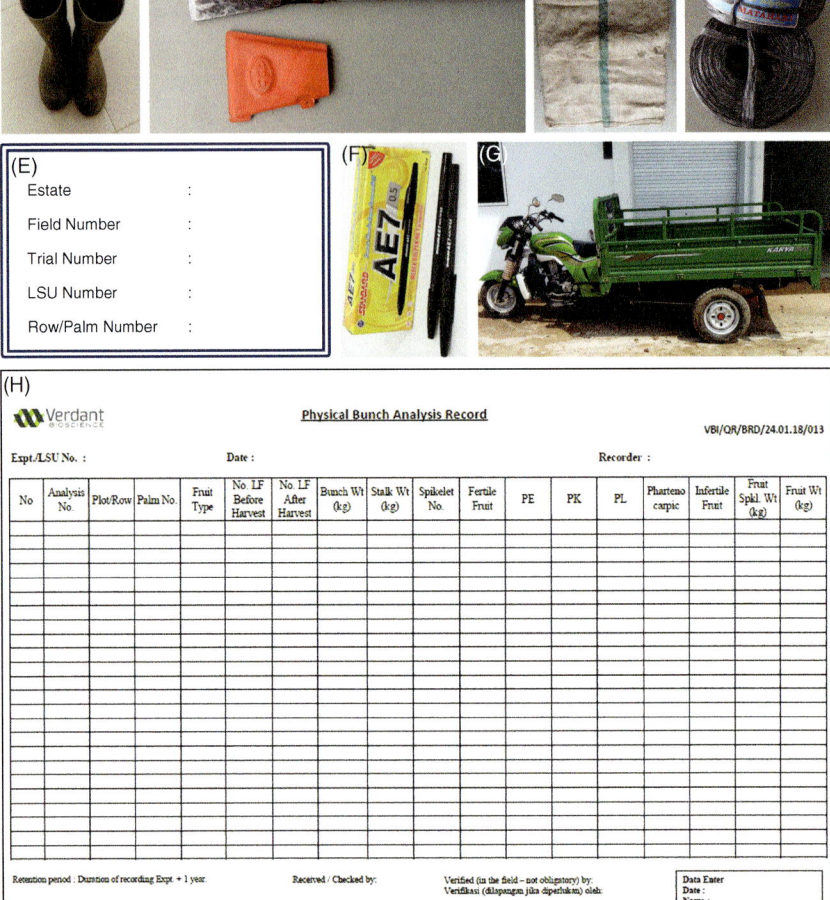

Fig. 3.7. Bunch sampling tools: A) Boots; B) Chisel; C) Gunny sack; D) Plastic rope; E) Label; F) Ballpoint pen; G) Three-wheel motorcycle; H) Physical bunch analysis record.

The steps in bunch sampling are summarized in Fig. 3.8.

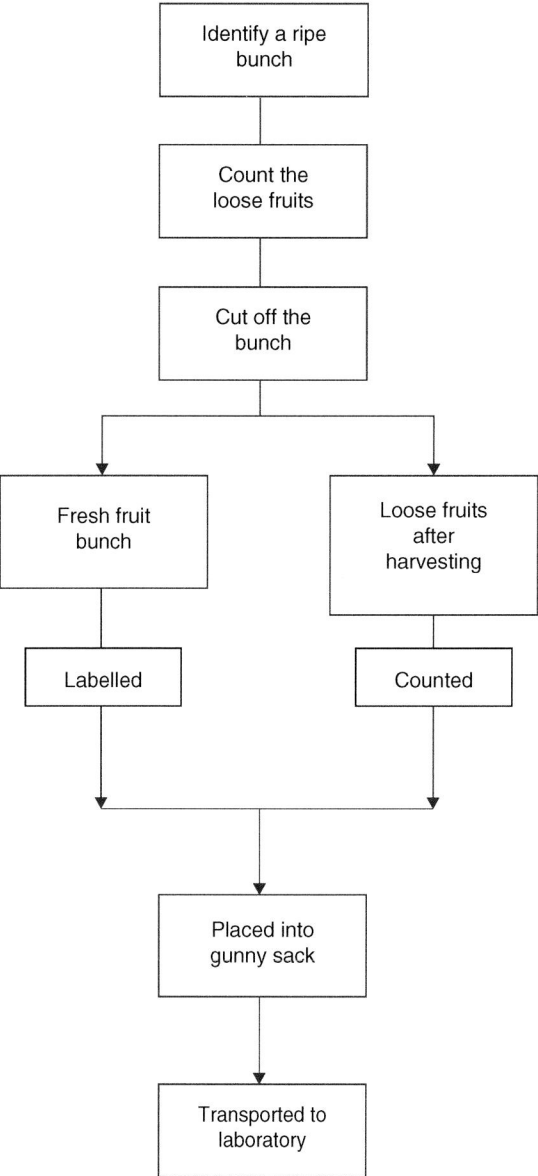

Fig. 3.8. Steps in ripe fruit bunch sampling.

Bunch Physical Analysis

4

Abstract

Fresh fruit bunches harvested from the field should be transferred to the physical analysis laboratory as soon as possible to determine: bunch sample weight; the fruit type (Dura, Tenera and Pisifera); stalk weight; total number of spikelets in the bunch; percentage of fertile fruits in the bunch (F/B); number of parthenocarpic fruits in the bunch; number of under-developed fruits in the bunch; fruit set (FS); fruit weight (FWT); percentage of mesocarp in the fruits (M/F); and percentage of mesocarp in the fruit check (M/FC). Spikelets are separated from the stalk (after an overnight fermentation process with ethephon solution) by chopping with an axe. A random sample of spikelets (according to bunch weight) is taken, from which fruits are removed. The separated fruits are classified as either fertile, parthenocarpic or under-developed. A random sample of 30 fertile fruits without any bruising or damage is then taken for nut and oil analysis. But first the fruits should be scraped to clean off the mesocarp from the nut. The data required for calculating various parameters are recorded: fruit type; bunch weight; stalk weight; spikelet number; spikelet sample weight; fruit sample weight; 30 fruit sample weight; wet (fresh) mesocarp weight; and number of nuts.

4.1 Steps in Bunch Physical Analysis

Step 1

Drill a hole in the bunch stalk using an electric drill (diameter 10 mm and length 15 cm) for injection of 1% ethephon solution (inject 1 ml solution into each 1 kg bunch) or spray a solution around the bunch. The bunch is placed in a gunny sack and fermented overnight (Fig. 4.1).

Fig. 4.1. Step 1: Drilling a hole in bunch stalk for ethephon injection.

Step 2

Take the field label and record data on a form. Then take three fruits from the bunch sample, cut the fruits in half with a knife and observe the fruit type (shell thickness: Dura – thick-shelled, Tenera – thin-shelled or Pisifera – no shell, see Figs 1.1 and 4.2). Record the observations.

Fig. 4.2. Step 2: Observation of fruit type in cut fruits (Dura, thick-shelled, in this example).

Step 3

Weigh the bunch using a digital platform balance to determine bunch weight. Record the bunch weight (Fig. 4.3).

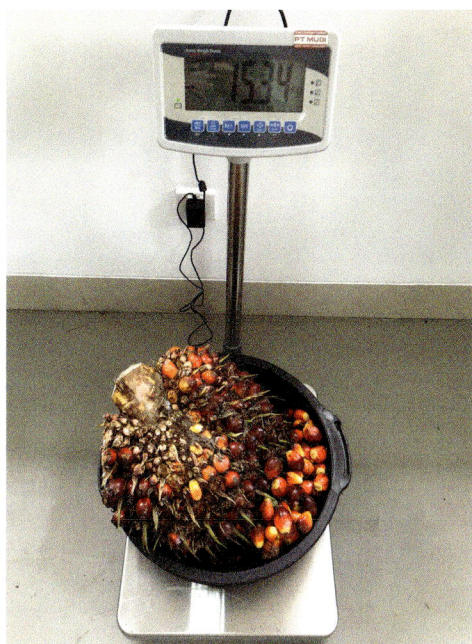

Fig. 4.3. Step 3: Weighing the bunch.

Step 4

Chop the bunch with an axe to separate spikelets from the stalk (note, this is a specialized task for which training is required). Chopping is conducted on a chopping table. Protective clothing, gloves and goggles must be worn (Fig. 4.4).

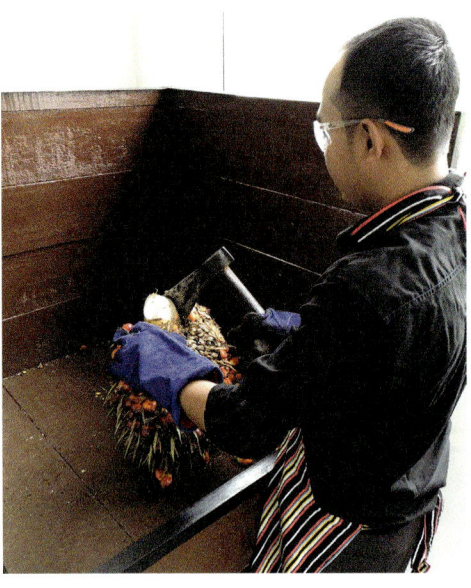

Fig. 4.4. Step 4: Chopping the bunch with an axe on a chopping table to separate and remove all the spikelets.

Step 5

Weigh the stalk using a digital platform balance. Record the stalk weight (Fig. 4.5).

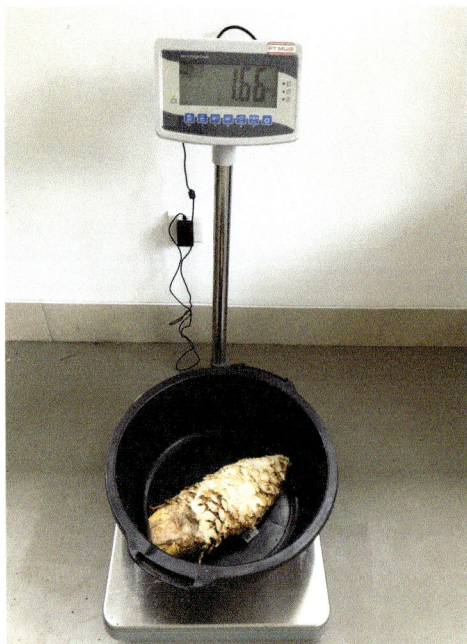

Fig. 4.5. Step 5: Weighing the stalk after removal of all the spikelets.

Step 6

Count the number of spikelets using a hand tally counter. The tally counter should be checked before use each day. If found inaccurate, discard and replace. Weigh the spikelets (Fig. 4.6).

Fig. 4.6. Step 6: Counting the spikelets.

Step 7

Take a specific amount (weight) of spikelets as determined by the bunch weight (Table 4.1). The spikelets are randomized by placing them on the chopping table and tipping them into four boxes (Fig. 4.7).

Table 4.1. Spikelet sample size as determined by bunch weight.

No.	Bunch weight (kg)	Percentage sampled	Number of boxes needed
1	<7.5 kg	100%	4
2	7.6–11.5 kg	75%	3
3	11.6–16.5 kg	50%	2
4	16.6–23.0 kg	37.5%	1.5
5	>23.0 kg	25%	1

Note: to get 1.5 boxes of samples, first take two boxes randomly from the original spikelet sample (50% sampling size). Then place the contents of the two boxes on the chopping table and drop them into four boxes. After that, take three of the four boxes (37.5% sampling size) with one box being discarded.

Fig. 4.7. Step 7: Random sampling the spikelets by dropping into boxes.

Step 8

Weigh the spikelet sample using a digital platform balance. Record the weight of the spikelets (Fig. 4.8).

Fig. 4.8. Step 8: Weighing the spikelets.

Step 9

Detach the fruits from the spikelets by hand-peeling or by using a knife. Then count the number of fertile, parthenocarpic and under-developed fruits. Parthenocarpic and under-developed fruit are removed from the fruit sample. Record the number of fertile fruits, parthenocarpic fruits and undeveloped fruits (Fig. 4.9).

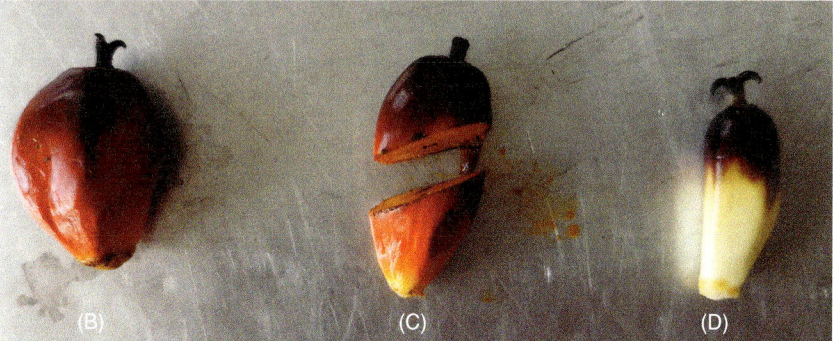

Fig. 4.9. Step 9: A) Detaching the fruits from the spikelets; B) Fertile fruit; C) Parthenocarpic fruit (cut open to reveal the lack of a kernel); D) Under-developed fruit.

Step 10

Weigh the fruit sample using a digital platform balance. Record the fruit weight (Fig. 4.10).

Fig. 4.10. Step 10: Weighing the fruit sample.

Step 11

Take a sample of 30 fruits from a randomization box (Fig. 4.11). Refer to the next chapter of this manual (Chapter 5).

Fig. 4.11. Step 11: Take a sample of 30 fruits.

Step 12

Weigh the 30-fruit sample using an analytical balance. Before weighing the fruits, make sure the total fruit number is 30. Record the fruit weight (Fig. 4.12).

Fig. 4.12. Step 12: Weighing the 30-fruit sample.

Step 13

Separate the mesocarp from the nut using a knife and produce a bulked wet (fresh) mesocarp sample. The scraped mesocarp and nuts are placed into separate aluminium trays for oven drying (Fig. 4.13).

Fig. 4.13. Step 13: Scraping the fruits with a knife to remove the mesocarp from the nut (note, this is a skilled task for which training is needed).

Step 14

Weigh the tray with the scraped wet mesocarp sample using an analytical balance. For further steps in oil analysis see Chapter 7. Record the wet mesocarp weight (Fig. 4.14).

Fig. 4.14. Step 14: Weighing the scraped wet mesocarp sample.

Step 15

Count the number of nuts (the total number of nuts should be 30). For further steps in nut analysis see Chapter 6. Record the number of nuts (Fig. 4.15).

Fig. 4.15. Step 15: Counting the nuts.

4.2 Tools and Equipment

- Electrical drill – used to drill a channel (2,800 rpm, Ø10mm and 12cm) into the bunch stalk for ethephon injection.
- Syringe, 10 ml – used to inject ethephon solution into the stalk.
- Gunny sack – used for storing the bunch during the fermentation step.
- Axe – used to chop the bunch into spikelets.
- Chopping table – used as a base for bunch chopping.
- Randomization box – used to randomize fruit samples.
- Digital platform balance, accuracy 0.01 kg, max 60 kg – used for weighing the bunch, stalk and spikelets.
- Tally counter – used to count spikelets, fruits and nuts.
- Analytical balance, accuracy 0.01 g, max 3,200 g – used for weighing fruits and nuts.
- Cloth gloves – used for handling the bunch during the chopping process.
- Knife – used to scrape mesocarp from the nut.
- Aluminium tray – used for storing fruit and nut samples.
- Goggles – used for eye protection during bunch chopping.
- Protective clothing – used to protect the body from shrapnel during the chopping process.
- Recording form – used to record data on fruit type, bunch weight, stalk weight, spikelet number, spikelet sample weight, fertile fruit number,

fertile fruit weight, parthenocarpic fruit number, under-developed fruit number, spikelet sample weight, fruit weight, 30-fruit sample weight, wet fresh mesocarp weight and number of nuts.
- Plastic container, 15 L – used for weighing accessories and samples.

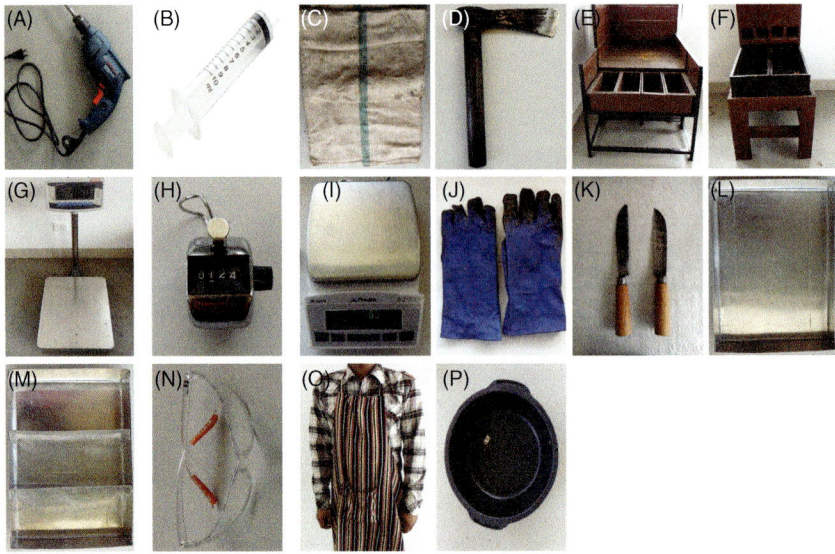

Fig. 4.16. Bunch physical analysis tools and equipment: A) Electrical drill; B) Syringe; C) Gunny sack; D) Axe; E) Chopping table; F) Randomization box; G) Digital platform; H) Tally counter; I) Analytical balance; J) Cloth gloves; K) Knife; L) Fruit tray; M) Nut tray; N) Goggles; O) Protective clothing; P) Plastic container; Q) Oil analysis record. Please also see physical bunch analysis record (Fig. 3.7H).

4.3 Materials

Ethephon (2-Chloroethylphosphonic acid) – used to induce fermentation of the bunch. Ethephon is a chemical that functions as a synthetic plant growth regulator.

Fig. 4.17. Bunch physical analysis material: ethephon, used to ferment the bunch which helps in loosening and detaching spikelets and fruits.

Formulae

1. Percentage of fertile fruits in the bunch (F/B)
   Percentage of fertile fruits in the bunch (F/B) = ((((bunch weight − stalk weight) ÷ fruit spikelet weight) x fruit weight) ÷ bunch weight) x 100
2. Fruit set (FS)
   Fruit set (FS) = (fertile fruits ÷ (fertile fruits + parthenocarpic fruits + under-developed fruits)) x 100
3. Fruit weight (FWT)
   Fruit weight (FWT) = 30 fruit sample weight ÷ 30
4. Percentage of mesocarp in the fruits (M/F)
   Percentage of mesocarp in the fruits (M/F) = ((30 fruit sample weight − nut weight) ÷ 30 fruit sample weight) x 100
5. Percentage of mesocarp in the fruit check (M/FC)
   Percentage of mesocarp in the fruit check (M/FC) = ((tray and fresh mesocarp weight − tray weight) ÷ (30 fruit sample weight − nut weight)) x 100

See Junaidah *et al.* (2011) for similar formulae.

The steps in physical bunch analysis are summarized in the schematic below.

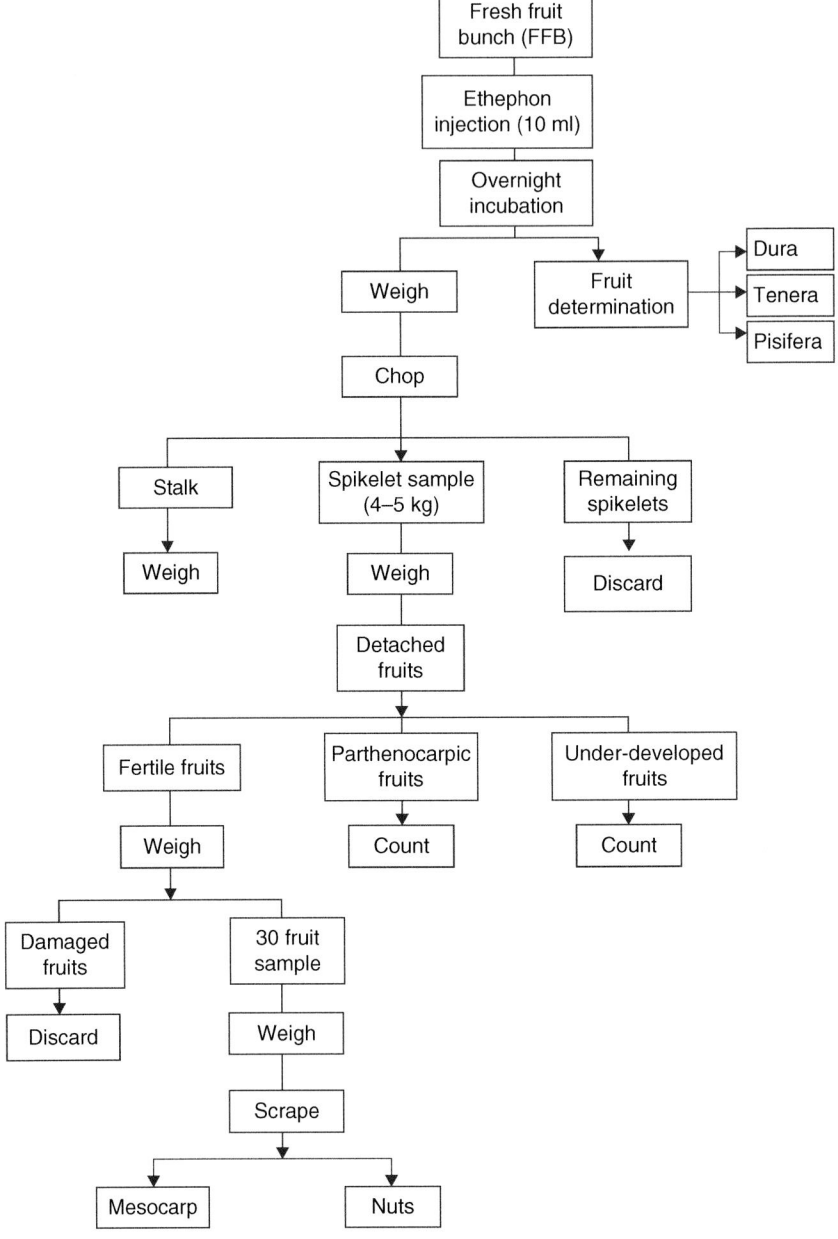

Fig. 4.18. Steps in bunch physical analysis.

Reference

Junaidah, J., Kushairi, A., Jones, B., Kho, L.E., Isa, Z.A. and Rusmin, J. (2011) Innovation for oil extraction method using NMR in bunch analysis. *International Seminar on Breeding for Sustainability in Oil Palm*, 18 November, ISOPB & MPOB, Kuala Lumpur, Malaysia, pp. 1–18.

Fruit Sampling

Abstract

Both nut analysis and oil analysis require proper sampling. Fertile fruits without any damage (bruising or chopped) are sampled using a randomization box. The randomization box divides the fruit sample into two portions (into two container boxes). Thirty fruits are taken randomly from each box and placed onto two separate trays. The weight deviation between these two replicated trays should be less than 5%.

5.1 Steps in Fruit Sampling

Step 1

Place all the fruit samples on top of the randomization box and remove all damaged (bruised and chopped) fruits (Fig. 5.1).

Fig. 5.1. Step 1: Fruit samples placed into randomization box.

Step 2

Open the floor of the randomization box to allow fruits to fall down randomly into two collection boxes (Fig. 5.2).

Fig. 5.2. Step 2: Open the floor of randomization box to allow fruit to fall into collecting boxes.

Step 3

Take 30 random fruits from each of the two boxes and place into two separate trays (Fig. 5.3).

Fig. 5.3. Step 3: Take 30 fruit randomly from each box.

Step 4

Count/check the number of fruits taken using a tally counter (the fruit number should be 30 per tray) (Fig. 5.4).

Fig. 5.4. Step 4: 30 counted fruits per tray.

Step 5

Weigh the 30-fruit sample in each tray using an analytical balance. Record the fruit sample weight. Note: the weight difference between the two trays should be less than 5%. If the difference is higher than 5%, repeat from Step 3 (Fig. 5.5).

Fig. 5.5. Step 5: Weighing 30-fruit sample.

5.2 Tools and Equipment

- Randomization box – used for sampling fruits randomly.
- Tally counter – used to count fruits and nuts.
- Analytical balance, accuracy 0.01 g, max 3,200 g – used to weigh fruits.
- Aluminium tray – used to store fruits.
- Recording form – used for recording the fruit number and 30-fruit sample weight.

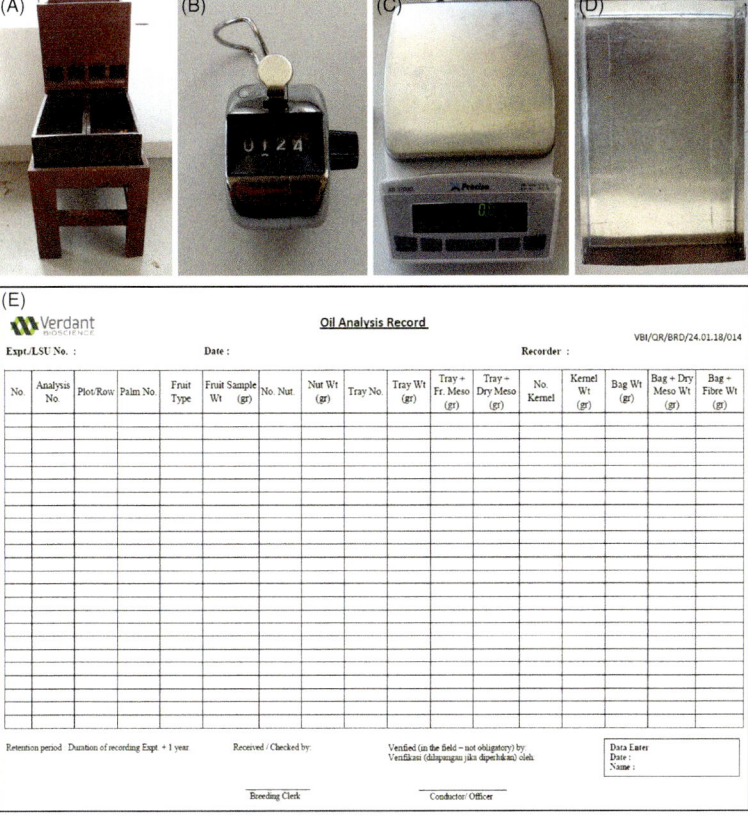

Fig. 5.6. Fruit sampling tools and equipment: A) Randomization boxes; B) Tally counter; C) Analytical balance; D) Aluminium tray; E) Recording form.

The steps in fruit sampling are summarized in the schematic below.

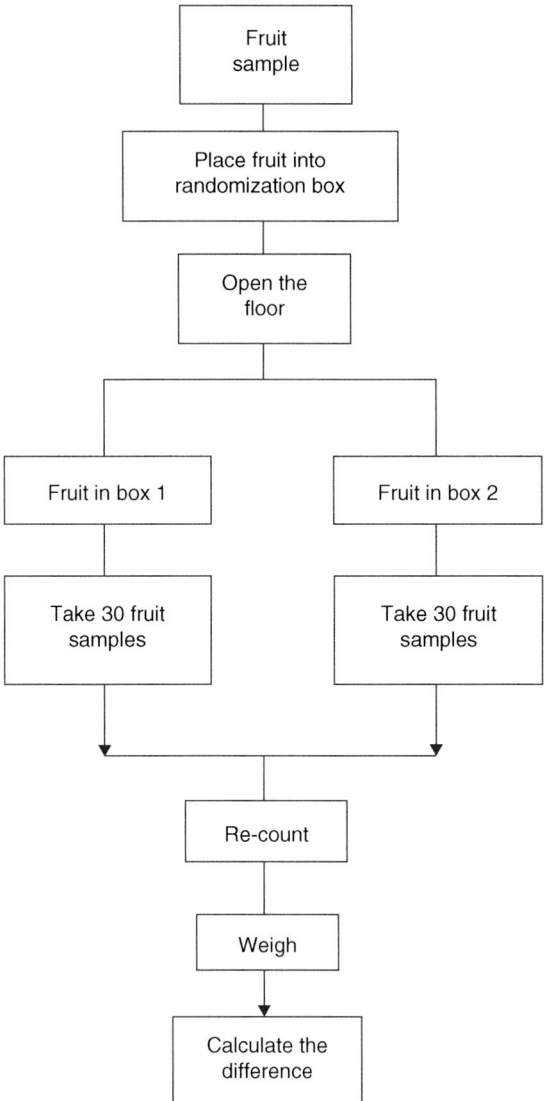

Fig. 5.7. Steps in providing a random sample of 30 fruits.

Nut Analysis

Abstract

Nut analysis provides data on: percentage of shell in the fruit (S/F); shell to kernel (S/K); percentage of kernel in the fruit (K/F); kernel weight (KW); nut weight (NW); kernel to nut (K/N); and percentage of kernel in the bunch (K/B). Nuts should be oven dried at 60–105°C for seven to nine hours, followed by cracking the nut shell to release the kernel.

6.1 Steps in Nut Analysis

Step 1

Place the sample of 30 fresh nuts (see Step 15, Chapter 4) onto an aluminium tray and weigh the nuts using an analytical balance. Record the nut sample weight (Fig. 6.1).

Fig. 6.1. Step 1: Weighing the nuts.

Step 2

Dry in an oven at 60–105°C for seven to nine hours (Fig. 6.2).

Fig. 6.2. Step 2: Drying the nuts in an oven.

Step 3

Remove the nuts from the oven and let them cool at room temperature. Crack the nut shell open with a hammer to release the kernel (note: this is a skill that requires training) (Fig. 6.3).

Fig. 6.3. Step 3: Cracking open the nut shell.

Step 4

Count the number of kernels. Record the kernel number (Fig. 6.4).

Fig. 6.4. Step 4: Counting the kernels.

Step 5

Weigh the kernels using an analytical balance. Record the kernel sample weight (Fig. 6.5).

Fig. 6.5. Step 5: Weighing the kernels.

6.2 Tools and Equipment

- Oven, temperature up to 105°C, fan and circulation – used to dry nuts.
- Analytical balance, accuracy 0.01 g, max 3,200 g – used for weighing nuts.
- Aluminium tray – used for storing nuts.
- Hammer – used for cracking open the nut shell.
- Recording form – used for recording nut weights, kernel weight, kernel number and kernel weight.

Formulae

1. Percentage of shell in the fruit (S/F)
   Percentage of shell in the fruit (S/F) = ((nut weight − kernel weight) ÷ 30 fruit sample weight) x 100
2. Shell to kernel (S/K)
   Shell to kernel (S/K) = (nut weight − kernel weight) ÷ kernel weight
3. Percentage of kernel in the fruit (K/F)
   Percentage of kernel in the fruit (K/F) = (kernel weight ÷ 30 fruit sample weight) x 100
4. Kernel to nut (K/N)
   Kernel to nut (K/N) = number of kernel ÷ number of nut
5. Percentage of kernel in the bunch (K/B)
   Percentage of kernel in the bunch (K/B) = (K/F x F/B) ÷ 100

See Junaidah *et al.* (2011) for similar formulae.

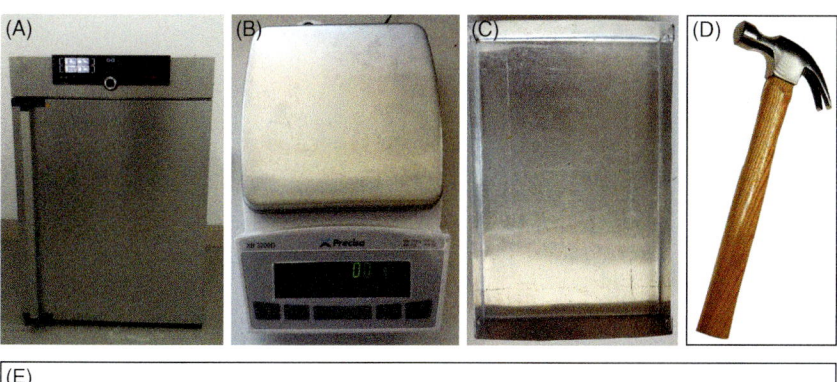

Fig. 6.6. Nut sampling tools and equipment: A) Oven; B) Analytical balance; C) Aluminium tray; D) Hammer; E) Recording form.

The steps in nut analysis are summarized in the schematic below.

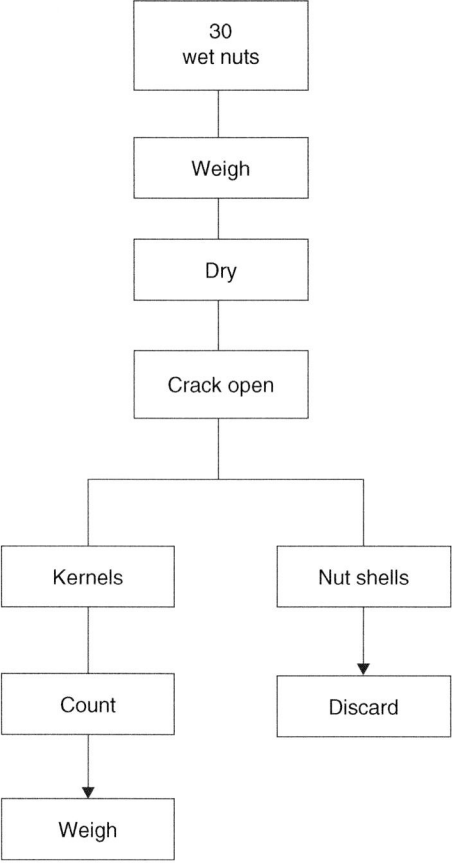

Fig. 6.7. Schematic of steps in nut analysis.

Reference

Junaidah, J., Kushairi, A., Jones, B., Kho, L.E., Isa, Z.A. and Rusmin, J. (2011) Innovation for oil extraction method using NMR in bunch analysis. *International Seminar on Breeding for Sustainability in Oil Palm*, 18 November, ISOPB & MPOB, Kuala Lumpur, Malaysia, pp. 1–18.

Oil Analysis

7

Abstract

Oil analysis of the mesocarp is determined by extracting the oil from dried mesocarp samples. Before the extraction step is conducted, the wet (fresh) mesocarp sample needs to be dried. Conventional drying of wet mesocarp is done in an oven. Oil is removed by refluxing with n-hexane in a soxhlet apparatus. Once oil extraction is completed, oil yield can be calculated by the difference of mesocarp weight before and after extraction. Oil analysis will obtain the percentage of dry mesocarp to fruit (DM/F); dry mesocarp to wet mesocarp (DM/WM); percentage of oil in dry mesocarp (O/DM); percentage of oil in wet mesocarp (O/WM); and percentage of mesocarp oil in the bunch (O/B).

7.1 Steps in Oil Analysis

Step 1

Dry wet (fresh) mesocarp (see Step 14, Chapter 4) in an oven for 24 hours at 60–105°C (Fig. 7.1).

Fig. 7.1. Step 1: Drying the fresh mesocarp in an oven.

Step 2

Take the tray from the oven using insulated gloves and let it cool at room temperature. Weigh the tray plus the dry mesocarp using an analytical balance. Make sure the sample weight is constant during weighing. Record the tray plus the dry mesocarp weight (Fig. 7.2).

Fig. 7.2. Step 2: Weighing the dry mesocarp.

Step 3

Take approximately 30 g of dry mesocarp and grind in a blender (Fig. 7.3).

Fig. 7.3. Step 3: Grinding the dry mesocarp in a blender.

Step 4

Take out the ground mesocarp from the blender and put it into a round aluminium sieve dish (Fig. 7.4).

Fig. 7.4. Step 4: Place the ground dry mesocarp into a round aluminium sieve dish.

Step 5

Sieve the ground mesocarp using a sieve with aperture size approximately 2 mm (Fig. 7.5).

Fig. 7.5. Step 5: Sieve the ground dry mesocarp.

Step 6

Place sieved mesocarp into a glass beaker and dry in an oven for 30 minutes at 60–105°C (Fig. 7.6).

Fig. 7.6. Step 6: A) Store the ground dry mesocarp in a glass beaker; B) Dry the ground dry mesocarp in an oven.

Step 7

Make an envelope from Whatman filter paper approximately 8 x 6 cm. Then weigh the envelope. Record the envelope weight (Fig. 7.7).

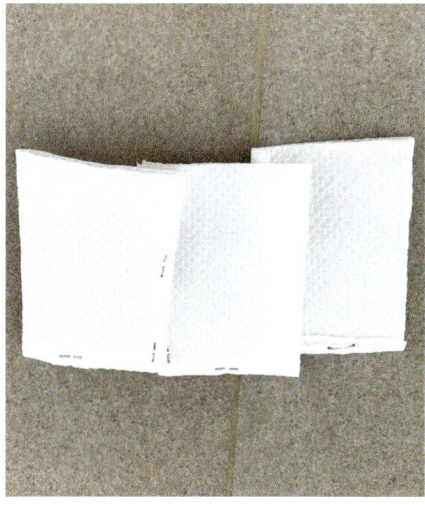

Fig. 7.7. Step 7: Envelopes made from stapled Whatman filter paper.

Step 8

Take approximately 8 g of ground mesocarp sample and place inside an envelope, staple to close and weigh the envelope plus the mesocarp sample. Write a label on the envelope using a graphite pencil. The analysis number should be written on the envelope for sample identification. Record the envelope plus mesocarp sample weight (Fig. 7.8).

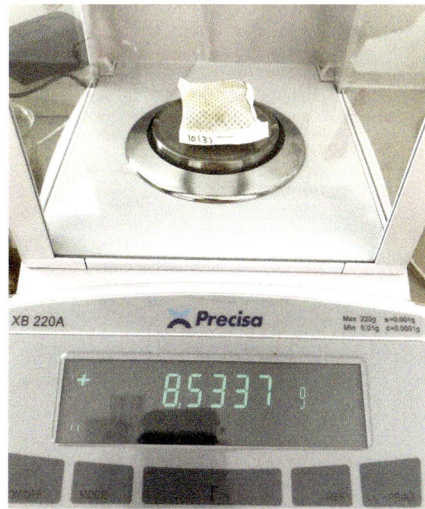

Fig. 7.8. Step 8: Weigh the envelope containing the mesocarp sample.

Step 9

Keep the enveloped mesocarp samples in a desiccator prior to the oil extraction step (Fig. 7.9).

Fig. 7.9. Step 9: Keep the enveloped mesocarp samples in desiccator prior to oil extraction if not used immediately.

Step 10

Arrange bags containing mesocarp samples into the soxhlet extractor system and then attach the heating mantle, boiling flask, soxhlet and condenser. One soxhlet can be filled with approximately 90 envelopes. Extract with n-hexane for 24 hours or until the solvent becomes clear. Extraction should be carried out under an exhaust system. Note: n-hexane is a hazardous chemical and care should be taken, following safety, health and environment guidance (see also Chapter 2 on storage, use and disposal of hazardous chemicals) (Fig. 7.10).

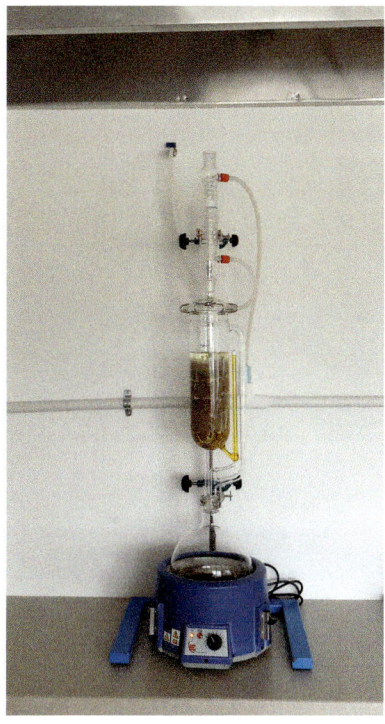

Fig. 7.10. Step 10: Extraction of enveloped dry mesocarp samples using a soxhlet extraction system.

Step 11

After extraction is complete, remove all envelopes from the soxhlet sytem and allow them to dry at room temperature for approximately 30 minutes (Fig. 7.11).

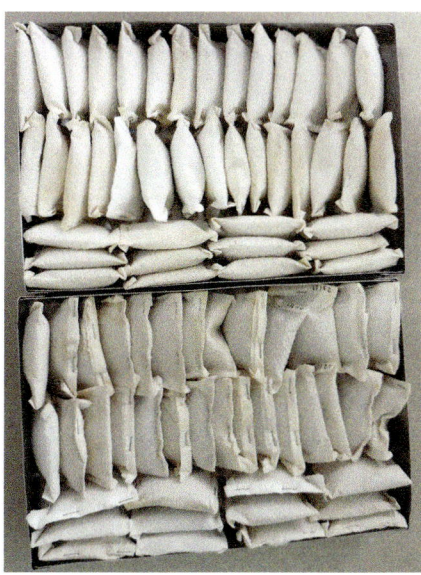

Fig. 7.11. Step 11: Dry the oil-extracted mesocarp samples.

Step 12

Weigh the envelope containing the mesocarp sample after extraction. Make sure the sample weight is constant during weighing. Record the envelope and mesocarp weight after drying (Fig. 7.12).

Fig. 7.12. Step 12: Weigh the envelope containing the mesocarp sample after extraction.

7.2 Tools and Equipment

- Oven, temperature up to 105°C, with a circulation fan – used for drying mesocarp samples.
- Analytical balance, accuracy 0.01 g, max 3,200 g – used to weigh the dry mesocarp.
- Analytical balance, accuracy 0.0001 g, max 220 g – used for weighing the dry mesocarp, before and after extraction.
- Aluminium tray – used for storing the mesocarp samples.
- Blender machine – used for grinding the dry mesocarp.
- Sieve, aperture size approximately 2 mm – used for sieving the ground mesocarp.
- Round aluminium dish – used for storing the ground mesocarp.
- Glass beaker 100 ml – used to store the ground mesocarp.
- Whatman filter paper – used to make envelopes for the ground mesocarp samples for extraction purposes.
- Ruler – used for measuring the Whatman envelope size.
- Staple, No. 10 – used for stapling the Whatman filter paper envelope.

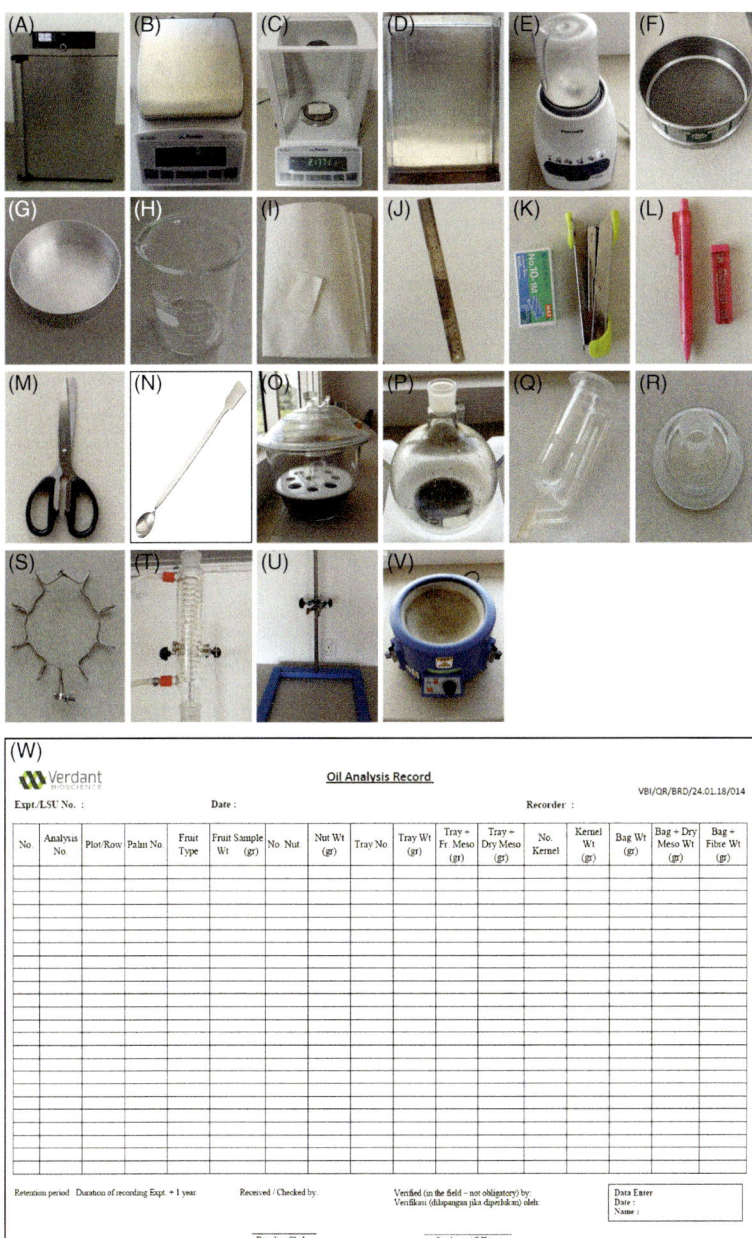

Fig. 7.13. Oil analysis tools and equipment: A) Oven; B) Analytical balance 3,200 g; C) Analytical balance 220 g; D) Aluminium tray; E) Blender; F) Sieve; G) Round aluminium dish; H) Glass beaker; I) Whatman filter paper; J) Ruler; K) Staple; L) Graphite pencil; M) Scissors; N) Spatula; O) Desiccator; P) Boiling flask; Q) Soxhlet extractor; R) Soxhlet extractor lid; S) Soxhlet extractor clamp: T) Condenser; U) Stand and clamp; V) Heating mantle; W) Recording form.

- Graphite pencil – used for labelling the Whatman envelope.
- Scissors – used for cutting the Whatman filter paper.
- Spatula – used to transfer the mesocarp sample into the Whatman envelope.
- Desiccator – used for storing the samples in Whatman envelopes prior to oil extraction.
- Soxhlet extractor system (includes round-bottom flask capacity 5 L, soxhlet extractor capacity 2 L, soxhlet extractor lid, soxhlet extractor clamp, stand and clamp, heating mantle and water cooling circulation system) – used to extract oil from dry mesocarp samples.
- Recording form – used to record the dry mesocarp sample weight, envelope weight, envelope + mesocarp sample weight before extraction and bag plus mesocarp sample weight after extraction.

7.3 Materials

Silica gel blue granules – used for absorbing moisture in the desiccator; n-hexane, technical grade – used to extract oil from the dry mesocarp (Fig. 7.14).

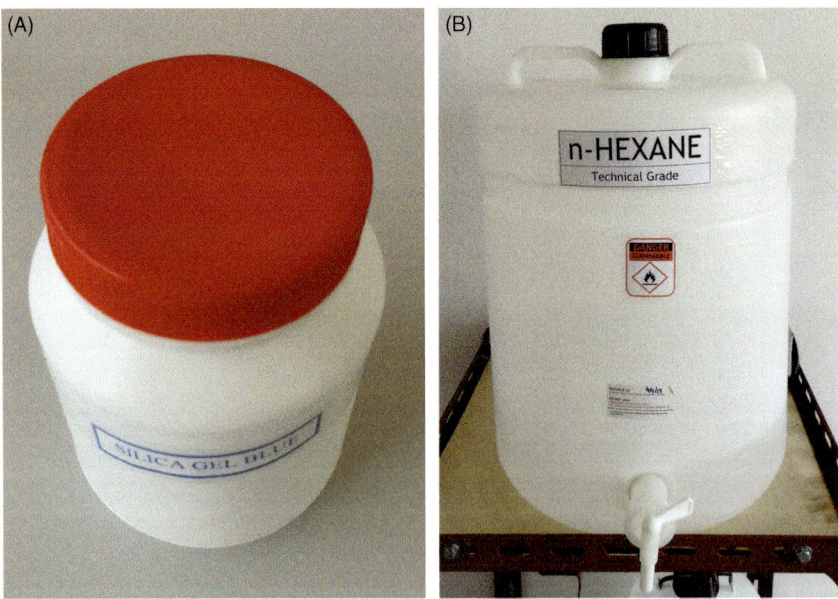

Fig. 7.14. Fruit sampling materials: A) Silica gel; B) n-hexane.

Formulae

1. Percentage of dry mesocarp to fruit (DM/F)
   Percentage of dry mesocarp to fruit (DM/F) = ((tray and dry mesocarp weight − tray weight) ÷ 30 fruit sample weight) x 100
2. Dry mesocarp to wet mesocarp (DM/WM)
   Dry mesocarp to wet mesocarp (DM/WM) = (tray and dry mesocarp weight − tray weight) ÷ (tray and fresh mesocarp weight − tray weight)
3. Percentage of oil in dry mesocarp (O/DM)
   Percentage of oil in dry mesocarp (O/DM) = ((envelope and dry mesocarp weight − Envelope and fibre weight) ÷ (envelope and dry mesocarp weight − envelope weight)) x 100
4. Percentage of oil in wet mesocarp (O/WM)
   Percentage of oil in wet mesocarp (O/WM) = ((tray and dry mesocarp weight − tray weight) ÷ (30 fruit sample weight − nut weight)) x O/DM x 100
5. Percentage of mesocarp oil in the bunch (O/B)
   Percentage of mesocarp oil in the bunch (O/B) = (F/B x M/F x O/WM) ÷ 10000

See Junaidah *et al*. (2011) for similar formulae.

The steps in oil analysis are summarized in the schematic below.

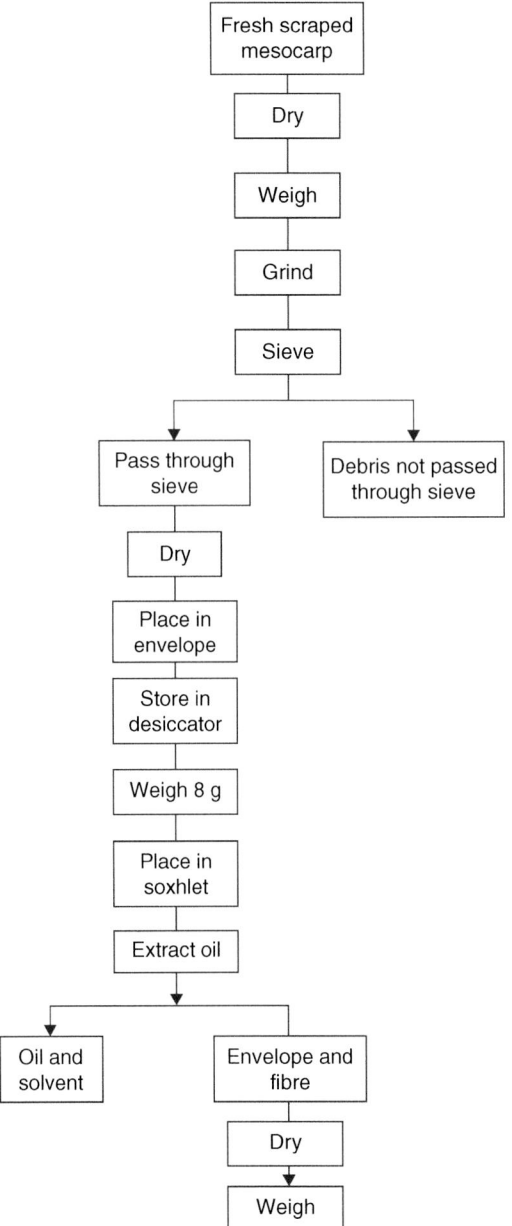

Fig. 7.15. Schematic of steps in oil analysis.

Reference

Junaidah, J., Kushairi, A., Jones, B., Kho, L.E., Isa, Z.A. and Rusmin, J. (2011) Innovation for oil extraction method using NMR in bunch analysis. *International Seminar on Breeding for Sustainability in Oil Palm*, 18 November, ISOPB & MPOB, Kuala Lumpur, Malaysia, pp. 1–18.

Recording, Calculations and Data Checks

Abstract

The records of analyses in the previous chapters are calculated by defined formulae to obtain data on: bunch components; percentage of fertile fruits in the bunch (F/B); fruit set (FS); fruit weight (FWT); percentage of mesocarp in the fruits (M/F); percentage of mesocarp in the fruit check (M/FC); percentage of shell in the fruit (S/F); shell to kernel (S/K); percentage of kernel in the fruit (K/F); kernel to nut (K/N); percentage of kernel in the bunch (K/B); percentage of dry mesocarp to fruit (DM/F); dry mesocarp to wet mesocarp (DM/WM); percentage of oil in dry mesocarp (O/DM); percentage of oil in wet mesocarp (O/WM); and percentage of mesocarp oil in the bunch (O/B).

8.1 Recording

Data recorded in physical analysis, an example.

Table 8.1. Raw data of physical analysis.

No.	Parameters (unit)	Unit	Value	Abbv.
1	Number	–	2	–
2	Analysis number	–	75	–
3	Plot/row	–	21	–
4	Palm number	–	6	–
5	Fruit type	–	T	–
6	Number LF before harvest	pcs	1	A
7	Number LF after harvest	pcs	9	B
8	Bunch weight	kg	21.01	C
9	Stalk weight	kg	2.14	D
10	Number of spikelets	pcs	227	E

Continued

Table 8.1. Continued.

No.	Parameters (unit)	Unit	Value	Abbv.
11	Fertile fruit	pcs	926	F
12	Parthenocarpic	pcs	0	G
13	Infertile fruit	pcs	368	H
14	Fruit spikelet weight	kg	7.20	I
15	Fruit weight	kg	5.35	J

Data recorded on oil analysis.

Table 8.2. Raw oil analysis data.

No.	Parameters (unit)	Unit	Value	Abbv.
1	30-fruit sample weight	g	200.2	K
2	Number of nuts	pcs	30	L
3	Nut weight	g	49.8	M
4	Tray number	–	8	N
5	Tray weight	g	145.9	O
6	Tray and fresh mesocarp weight	g	288.7	P
7	Tray and dry mesocarp weight	g	249.9	Q
8	Number of kernels	pcs	43	R
9	Kernel weight	g	19.6	S
10	Envelope weight	g	0.4396	T
11	Envelope and dry mesocarp weight	g	8.5976	U
12	Envelope and fibre weight	g	2.3831	V

8.2 Calculations

Calculation of bunch components is shown below, using a worked example.

1. Percentage of fertile fruits in the bunch (F/B)
   Percentage of fertile fruits in the bunch (F/B) = ((((bunch weight − stalk weight) ÷ fruit spikelet weight) x fruit weight) ÷ bunch weight) x 100
   F/B = ((((C − D) ÷ I) x J) ÷ C) x 100
   F/B = ((((21.01 − 2.14) ÷ 7.20) x 5.35) ÷ 21.01) x 100
   F/B = 66.74%
2. Fruit set (FS)
   Fruit set (FS) = (fertile fruits ÷ (fertile fruits + parthenocarpic fruits + under-developed fruits)) x 100
   FS = (F ÷ (F + G + H) x 100
   FS = (926 ÷ (926 + 0 + 368) x 100
   FS = 71.56%

3. Fruit weight (FWT)
   Fruit weight (FWT) = 30 fruit sample weight ÷ 30
   FWT = K ÷ 30
   FWT = 200.2 ÷ 30
4. Percentage of mesocarp in the fruit (M/F)
   Percentage of mesocarp in the fruit (M/F) = ((30 fruit sample weight − nut weight) ÷ 30 fruit sample weight) x 100
   M/F = ((K − M) ÷ K) x 100
   M/F = ((200.2 − 49.8) ÷ 200.2) x 100
   M/F = 75.12%
5. Percentage of mesocarp in the fruit check (M/FC)
   Percentage of mesocarp in the fruit check (M/FC) = ((tray and fresh mesocarp weight − tray weight) ÷ (30 fruit sample weight − nut weight)) x 100
   M/FC = ((P − O) ÷ (K − M)) x 100
   M/FC = ((288.7 − 145.9) ÷ (200.2 − 49.8)) x 100
   M/FC = 94.95%
6. Percentage of shell in the fruit (S/F)
   Percentage of shell in the fruit (S/F) = ((nut weight − kernel weight) ÷ 30 fruit sample weight) x 100
   S/F = ((M − S) ÷ K) x 100
   S/F = ((49.8 − 19.6) ÷ 200.2) x 100
   S/F = 15.08%
7. Shell to kernel (S/K)
   Shell to kernel (S/K) = (nut weight − kernel weight) ÷ kernel weight
   S/K = (M − S) ÷ S
   S/K = (49.8 − 19.6) ÷ 19.6
   S/K = 1.54
8. Percentage of kernel in the fruit (K/F)
   Percentage of kernel in the fruit (K/F) = (kernel weight ÷ 30 fruit sample weight) x 100
   K/F = (S ÷ K) x 100
   K/F = (19.6 ÷ 200.2) x 100
   K/F = 9.79%
9. Kernel to nut (K/N)
   Kernel to nut (K/N) = number of kernel ÷ number of nut
   K/N = R ÷ L
   K/N = 43 ÷ 30
   K/N = 1.43
10. Percentage of kernel in the bunch (K/B)
    Percentage of kernel in the bunch (K/B) = (K/F x F/B) ÷ 100
    K/B = (K/F x F/B) ÷ 100
    K/B = (9.79% x 66.74%) ÷ 100
    K/B = 6.53%

11. Percentage of dry mesocarp to fruit (DM/F)
    Percentage of dry mesocarp to fruit (DM/F) = ((tray and dry mesocarp weight − tray weight) ÷ 30 fruit sample weight) x 100
    DM/F = ((Q − O) ÷ K) x 100
    DM/F = ((249.9 − 145.9) ÷ 200.2) x 100
    DM/F = 51.95%
12. Dry mesocarp to wet mesocarp (DM/WM)
    Dry mesocarp to wet mesocarp (DM/WM) = (tray and dry mesocarp weight − tray weight) ÷ (tray and fresh mesocarp weight − tray weight)
    DM/WM = (Q − O) ÷ (P − O)
    DM/WM = (249.9 − 145.9) ÷ (288.7 − 145.9)
    DM/WM = 0.728
13. Percentage of oil in dry mesocarp (O/DM)
    Percentage of oil in dry mesocarp (O/DM) = ((envelope and dry mesocarp weight − envelope and fibre weight) ÷ (envelope and dry mesocarp weight − envelope weight)) x 100
    O/DM = ((U − V) ÷ (U − T)) x 100
    O/DM = ((8.5976 − 2.3831) ÷ (8.5976 − 0.4396) x 100
    O/DM = 76.18%
14. Percentage of oil in wet mesocarp (O/WM)
    Percentage of oil in wet mesocarp (O/WM) = ((tray and dry mesocarp weight − tray weight) ÷ (30 fruit sample weight − nut weight)) x O/DM x 100
    O/WM = ((Q − O) ÷ (K − M)) x O/DM x 100
    O/WM = ((249.9 − 145.9) ÷ (200.2 − 49.8) x 76.18% x 100
    O/WM = 52.68%
15. Percentage of mesocarp oil in the bunch (O/B)
    Percentage of mesocarp oil in the bunch (O/B) = (F/B x M/F x O/WM) ÷ 10,000
    O/B = (F/B x M/F x O/WM) ÷ 10000
    O/B = 66.74% x 75.12% x 52.68% ÷ 10,000
    O/B = 26.41%

The report of bunch components.

Table 8.3. Report of bunch components.

No.	Bunch component	Abbv. (unit)	Value
1	Percentage of fertile fruits in the bunch (F/B)	F/B (%)	66.74%
2	Fruit set (FS)	FS (%)	71.56%
3	Fruit weight (FWT)	FWT	6.67
4	Percentage of mesocarp in the fruits (M/F)	M/F (%)	75.12%
5	Percentage of mesocarp in the fruit check (M/FC)	M/FC (%)	94.95%
6	Percentage of shell in the fruit (S/F)	S/F (%)	15.08%
7	Shell to kernel (S/K)	S/K (%)	1.54
8	Percentage of kernel in the fruit (K/F)	K/F (%)	9.79%
9	Kernel to nut (K/N)	K/N	1.43

Continued

Table 8.3. Continued.

No.	Bunch component	Abbv. (unit)	Value
10	Percentage of kernel in the bunch (K/B)	K/B (%)	6.53%
11	Percentage of dry mesocarp to fruit (DM/F)	DM/F (%)	51.95%
12	Dry mesocarp to wet mesocarp (DM/WM)	DM/WM	0.728
13	Percentage of oil in dry mesocarp (O/DM)	O/DM (%)	76.18%
14	Percentage of oil in wet mesocarp (O/WM)	O/WM (%)	52.68%
15	Percentage of mesocarp oil in the bunch (O/B)	O/B (%)	26.41%

8.3 Tools and Equipment

Calculator – used for calculating bunch components.

Fig. 8.1. Fruit sampling tools – calculator.

8.4 Data Checks and Quality Control

Data checking and quality control of data are needed to screen for anomalies/outliers or out-of-range data that is apparent during processing.

There is much information in books and journals about formulae for bunch and oil analysis (e.g. Junaidah *et al.*, 2011; Corley and Tinker, 2015). Here we provide a simple standard value for data checking.

Table 8.4. Data checks for Dura and Tenera samples. Samples with data values outside these guide figures should be scrutinized very carefully and potentially eliminated.

No.	Parameters	Dura		Tenera	
		More than	Less than	More than	Less than
1	F/B	85%	50%	85%	50%
2	FW	33 g	4.5 g	25 g	4.5 g
3	M/F	74%	43%	94%	64%
4	M/FC	100	93	100	93
5	DM/F	60%	23%	80%	35%
6	DM/WM	0.785	0.4	0.785	0.4
7	O/DM	95%	60%	96.80%	56.00%
8	S/F	46%	19%	24.30%	2.00%

The criteria limits are imposed to make sure that the quality of data used are good and to preclude distortion of results from the use of anomalous data.

For data checking, during data entering and processing a formula is prepared using the criteria above to reveal outlier data and show the reason for the data being anomalous. The outlier may be because of mis-entry of data, or human/equipment error during analysis, or it may be that data are true values but outside the normal range.

A simple way to check the data is to create a graph to see the inner and outer limits of the results.

The samples which lie outside the data range need to be re-checked (Fig. 8.2).

Example graphs for Tenera samples.

Fig. 8.2. Bunch components for Tenera samples.

Example graphs for Dura samples.

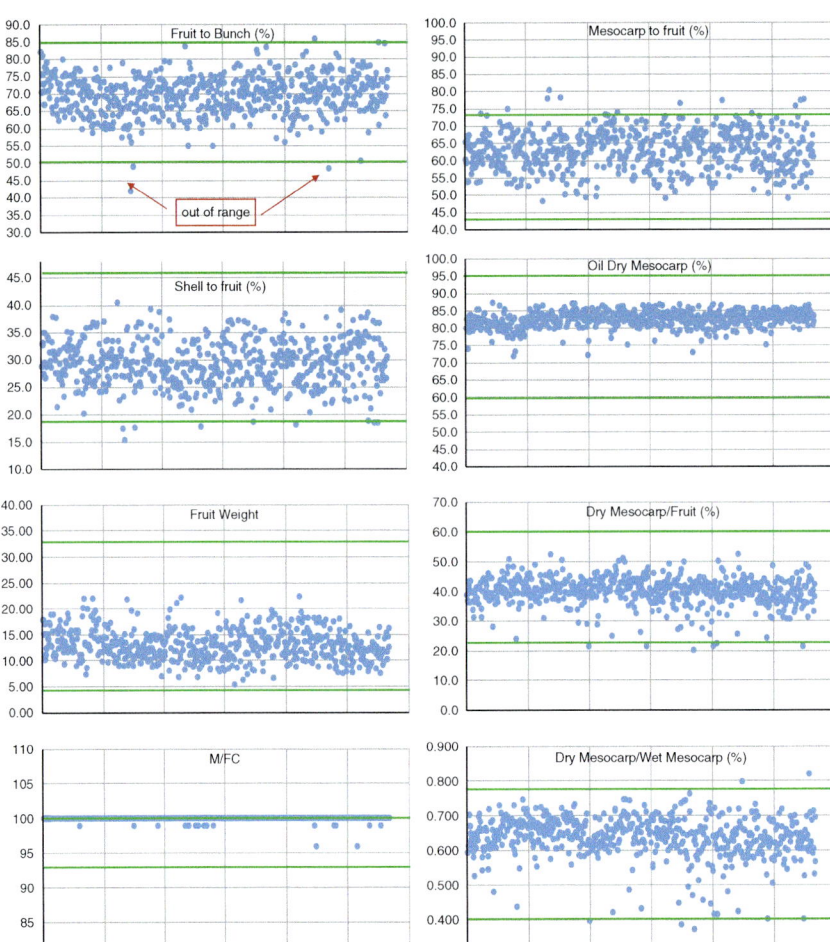

Fig. 8.3. Bunch components for Dura samples.

Produce quality control chart

Quality control charts are useful for monitoring the stability of bunch analysis processes and fluctuations in results (in a statistical context) over time. There are three types of line in the quality control graph:

Mean	Mean of the sample data
Inner control limit	twice of standard deviation for upper and lower limit
Outer control limit	three times of standard deviation for upper and lower limit

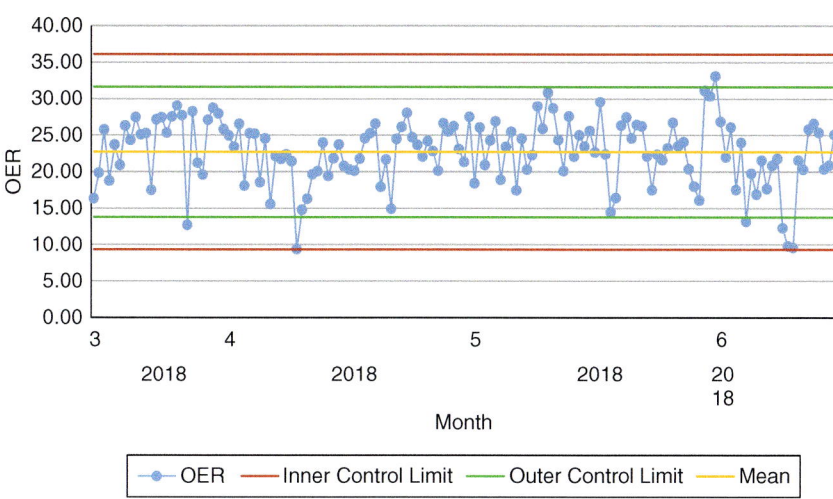

Fig. 8.4. Control chart of oil extraction rate (OER).

References

Corley, R.H.V. and Tinker, P.B. (2015) *The Oil Palm*, fifth edition. Wiley Blackwell, UK.

Junaidah, J., Kushairi, A., Jones, B., Kho, L.E., Isa, Z.A. and Rusmin, J. (2011) Innovation for oil extraction method using NMR in bunch analysis. *International Seminar on Breeding for Sustainability in Oil Palm*, 18 November, ISOPB & MPOB, Kuala Lumpur, Malaysia, pp. 1–18.

Index

Page numbers in **bold** type refer to figures and tables.

bunch sampling
    observations and records 22, 23
    procedure, steps 21–24, **27**
    tools and equipment 25, **26**
bunch weight 31, 33
bunch/yield analysis process
    example of data and calculations 75–76, 76–78, 78–79
    laboratory organisation 11, 11–13, **12**
    stages and methods 7–10, **10**, **13**

calculations
    data and quality control checks 79–82, **80**, **83**
    formulae 41, 56, 71
    worked examples from data 76–78
chemical hazards 18–19, 66
chopping (bunches) 31, **32**
colour, ripe/unripe fruit 6, **6**, 7, 21, **22**
crude palm oil (CPO) 4, 5, 7

data checking 79–80, **80**, **81**, **82**
desiccator, for sample storage 66, **66**, 70
dry mesocarp percentage in fruit (DM/F) 71, 78
drying
    ground dry mesocarp 64, **64**
    nut samples 53, 54, **54**
    soxhlet envelopes after extraction 67, **67**
    wet mesocarp samples 61, **62**
Dura oil palm type
    bunch component data, normal range **80**, **82**
    fruit characteristics 3, **4**, **30**
    oil composition and yield 4, **5**

equipment *see* tools and equipment
ethephon 8–9, 41, **41**
    injection into bunch stalk 29, **30**, 39
extraction (of oil from mesocarp) 10, 66–67, **67**

fatty acids
    composition of CPO and PKO 4–5
    free, content related to quality 7, 8
fermentation (after harvesting) 8–9, 29, 41
fertile fruits 35, **35**
    percentage in bunch (F/B) calculation 41, 76
filter paper envelopes **65**, 65–68
formulae (for calculations) 41, 56, 71
fruit sampling
    checking number of fruits 47, **47**
    fruit quality and sample size 9, 45
    random selection process 36, 45–47, **47**, **50**
    tools and equipment 48, **49**
    weight difference, replicate samples 45, 48
fruit set (FS) determination 35, 41, 76
fruit type (shell thickness) 4, 5, 30, **30**
fruit weight
    mean individual weight (FWT) 9, 41, 77
    total, 30-fruit samples 37, **37**, 48, **48**
    whole sample from bunch, fertile fruits 36, **36**

grinding, dry mesocarp 63, **63**

harvesting procedures 5–7, **8**, 17–18, 23, **23**
health and safety
    field 17–18
    laboratory 18–19
n-hexane 18–19, 66, 70, **70**

kernels
    kernel to nut (K/N) calculation 56, 77
    percentage in bunch (K/B) 56, 77
    percentage in fruit (K/F) 56, 77
    release from nut by cracking 54, **55**
    samples, number and weight 55, **55**, 56

labels 24, **24**, 30
laboratory operations
    health and safety 18–19
    set-up and work flow **11**, 11–13, **12**
    steps in bunch/oil analysis 8–10, **10**, 13
leaf sampling unit (LSU) 21, 24
loose fruits, number of
    after harvesting 23, **24**
    before harvesting 6–7, 22, **22**

mesocarp
    dry sample preparation and
        weights 61–66, **62**, **65**
    dry to wet (DM/WM) calculation 71, 78
    moisture content 7–8, 10
    percentage in fruit check (M/FC) 41, 77
    percentage in fruits (M/F),
        calculation 41, 77
    wet sample preparation and weight 37,
        37–38, **38**
    see also oil analysis
moisture content 7–8, 10

nut analysis
    data records and calculations 53, 56,
        76, 77
    number in samples 38, **39**
    process steps 9–10, 53–55, **58**
    separation from fruit 37, **37**
    tools and equipment 56, **57**
nut sample weight (30 nuts) 53

oil analysis
    example of raw data and
        calculations **76**, **78**, **79**
    process stages 10, 61–68, **72**

tools, equipment and materials 68, **69**,
    70, **70**
oil palm *(Elaeis guineensis)*
    crop production 3–4, 7
    fruit morphology 4, **5**
oil percentage in bunch (O/B) 71, 78
oil percentage in dry mesocarp (O/DM)
    71, 78
oil percentage in wet mesocarp (O/WM)
    71, 78
oil yield
    calculations, from oil analysis 61,
        71, 78
    of Dura, Pisifera and Tenera types
        3, **5**
    extraction rate (OER) quality
        control chart **83**
outliers (data) 79–80, **81**, **82**

palm kernel oil (PKO) 4–5, 7
parthenocarpic fruits 35, **35**
physical analysis of bunches
    data, records and calculations 29, 41,
        75–**76**, 76–77
    process steps 29–38, **42**
    tools and equipment 39–40, **40**
Pisifera oil palm type 3, **4**, **5**
protective clothing/equipment 17, 18, 31,
    39, **40**

quality control charts 82, **83**

randomization box **45**, 45–46, **46**, **47**
recording forms
    nut and oil analysis 56, **57**, 70, **76**
    physical bunch analysis **26**, 39–40,
        75–**76**
ripeness characteristics **6**, 6–7, 21, **22**

safety data sheets (SDS) 18
sampling
    bunch selection and harvesting
        21–24, **27**
    sample size related to bunch
        weight **33**
    spikelets, alternative methods 9,
        33–34, **34**
    see also fruit sampling
shell percentage in fruit (S/F) 56, 77
shell thickness *see* fruit type

shell to kernel (S/K) calculation 56, 77
sieving, dry mesocarp **63**, 63–64, **64**
soxhlet extraction system 10, 66–67, **67, 69**, 70
spikelets
    number, counting 33, **33**
    peeling fruits from 35, **35**
    removal from stalk (chopping) 31, **32**
    sampling methods 9, 33–34, **34**
    weight of sample 34, **34**
stalk weight 32
standard operating procedures (SOPs) 5, 7, 17, 19, 21

Tenera oil palm type
    bunch component data, normal range **80, 81**
    fruit characteristics 3, **4**, 5
    oil composition and yield 4, 5
tools and equipment
    bunch sampling 25, **26**
    fruit sampling 48, **49**
    nut analysis 56, **57**
    oil analysis 68, **69**, 70

physical analysis of bunches 39–40, **40**
transport of bunch samples 8, **8**, 24, **25**

under-developed fruits 35, **35**

vitamin content 4

waste, storage and disposal **12**
weighing
    fruit samples 36, **36**, 37, **37**, 48
    mesocarp after extraction 68, **68**
    nut/kernel samples 53, **54**, 55, **56**
    pre-extraction mesocarp samples 62, **62**, 65, **65**
    spikelet sample 34, **34**
    stalk, after spikelet removal 32, **32**
    whole bunch 31, **31**
wet mesocarp weight (WM) 38

yield components 7–8
    example of report, from data **78–79**

CABI – who we are and what we do

This book is published by **CABI**, an international not-for-profit organisation that improves people's lives worldwide by providing information and applying scientific expertise to solve problems in agriculture and the environment.

CABI is also a global publisher producing key scientific publications, including world renowned databases, as well as compendia, books, ebooks and full text electronic resources. We publish content in a wide range of subject areas including: agriculture and crop science / animal and veterinary sciences / ecology and conservation / environmental science / horticulture and plant sciences / human health, food science and nutrition / international development / leisure and tourism.

The profits from CABI's publishing activities enable us to work with farming communities around the world, supporting them as they battle with poor soil, invasive species and pests and diseases, to improve their livelihoods and help provide food for an ever growing population.

CABI is an international intergovernmental organisation, and we gratefully acknowledge the core financial support from our member countries (and lead agencies) including:

Discover more

To read more about CABI's work, please visit: **www.cabi.org**

Browse our books at: **www.cabi.org/bookshop**, or explore our online products at: **www.cabi.org/publishing-products**

Interested in writing for CABI? Find our author guidelines here:
www.cabi.org/publishing-products/information-for-authors/